NaturalEnergy

ALSO COAUTHORED BY **CAROL COLMAN:**

Shed Ten Years in Ten Weeks

The Melatonin Miracle: Nature's Age Reversing, Disease Fighting, Sex Enhancing Hormone

The Superhormone Promise

The Female Heart: The Truth About Women and Coronary Artery Disease

What Women Need to Know: From Headaches to Heart Disease and Everything in Between

Dr. Scott's Knee Book

The Lupus Handbook for Women

How to Prevent Miscarriage and Other Crises of Pregnancy

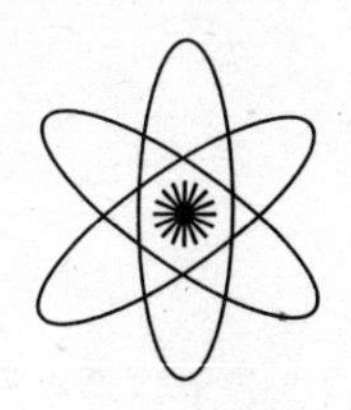

Natural Energy:

FROM TIRED TO TERRIFIC IN 10 DAYS

Erika Schwartz, M.D.,
& Carol Colman

VERMILION
London

1 3 5 7 9 10 8 6 4 2

Published by arrangement with GP Putnam's Sons, a member of Penguin Putnam Inc., New York

First published in the UK in 2000 by Vermilion, an imprint of Ebury Press, Random House, 20 Vauxhall Bridge Road, London SW1V 2SA

www.randomhouse.co.uk

Random House Australia (Pty) Limited
20 Alfred Street, Milsons Point, Sydney,
New South Wales 2061, Australia

Random House New Zealand Limited
18 Poland Road, Glenfield, Auckland 10, New Zealand

Random House (Pty) Limited
Endulini, 5a Jubilee Road, Parktown 2193, South Africa

The Random House Group Limited Reg. No. 954009

Papers used by Vermilion are natural, recyclable products made from wood grown in sustainable forests.

Printed and bound in Great Britain by
Biddles Ltd, Guildford and King's Lynn

A CIP catalogue record for this book is available from the British Library

ISBN 0 89 182730 2

There are two kinds of carnitine, L-carnitine and D-carnitine. Some studies suggest that D-carnitine may be toxic. Therefore, stick to products containing only L-carnitine.

To Lisa and Katie

—E.S.

To Michael and Josh

—C.C.

Contents

Acknowledgments

I am among those fortunate people who have been given the opportunity to share with the outside world the wisdom of twenty years of medical practice.

My father used to say, "Many are invited, few are chosen." I certainly would never have been among the chosen had I not been surrounded by a tremendous amount of trust, friendship, and support. It is the people who believe in me and the message I am carrying who deserve all the credit for this book.

Jeremy Katz, my editor—thank you for everything: for being my friend, my guiding light, a visionary who never tires or gives up, whose openminded approach to life makes it all sparkle and shine.

Carol Colman, my co-writer, all I have to say is—January 22—Lussardi's. The rest, as they say, is history. I love working with you, learning from you, sharing with you. Thank you.

Thank you to Ken Chandler, the editor of the *New York Post*—gently and unobtrusively you have guided me into a new world. I hope I serve you as well as your doctor as you do as my counsel.

Thank you to Karen Mayer, counsel to Penguin Putnam.

Jeanine Pirro, Westchester County DA, my friend of twenty years, my cheerleader and willing patient, how can I only thank you? Your faith in me makes me become a better doctor; your personal example, a better person.

To my friends and family—Elizabeth Armet, Daniel Armet, the Brates family, Odette Fodor, Sandy Taub, Joe Bova, Lisa and Bill Holmes, Venera Dumitrescu—thank you for listening. To my parents, Isac and Sylvia, thank you.

Thank you to our publisher, Phyllis Grann, to Marilyn Ducksworth, and certainly to my agent, Deborah Schneider, and Carol's agent, Richard Curtis.

Thank you to my partner in the practice of medicine for ten years, Andrew Fader, and to my office staff for their support.

To all my patients: you have taught me everything I know. You have made me who I am and your stories will now help others learn and take control of their own lives. I thank you for letting me into your lives.

Much thanks to all the researchers and scientists who so willingly shared their knowledge with us.

In particular, I would like to thank Dr. Claudio Cavazza and Dr. Menotti Calvini for their kind assistance. A special thanks to Dr. Eric Schon of Columbia University for his patient explanation of why old flies can't fly. His creative research provided the inspiration for that chapter.

I would also like to thank Elizabath Himelfarb, Jo Sgammato, Wende Fazio, and Adam Silverman for all their help.

Natural Energy

Chapter One

IT'S ALL ABOUT **ENERGY**

Picture this.

You wake up even before your alarm clock tells you to, yet you feel fully refreshed, recharged, and ready to take on the day. There are so many things you have to do . . . and you can't wait to get started. In the morning you have a new client meeting—you'll be smart and charming. At noon you're meeting friends at the gym for forty-five minutes of exercise and stretching. Then you have some quarterly reports to get out and some important phone calls to return. After work you'll be taking the kids to soccer practice and making them dinner—early, because you've arranged for a sitter so that you and your husband can try that new southwestern restaurant and take in a movie.

Is this a dream? Are your mornings more like this?

The alarm clock screams. You hit the snooze alarm and pull the bedcovers over your head, trying to squeeze in a few more minutes of precious sleep even as you begin to worry about what you didn't get done yesterday and all the piles facing you today. You drag yourself out of bed, exhausted and seriously considering calling in sick. After a shower and several cups of coffee you manage to get dressed and get the kids off to school, but one glance in the mirror reveals that you look as wiped out as you feel. You have an important new client meeting in the morning and you nearly nod off twice in the midst of it. At eleven you're feeling low-energy, so you have half a Danish and another cup of coffee, even though you're not really hungry. At lunch you eat a sandwich at your desk so that you can try to get some of yesterday's work done, but by two you're desperate for a nap. At seven all you want to do is put the kids to sleep and crawl back into bed. . . .

The second scenario probably seems a little too familiar. Chances are you often find yourself thinking, "I can't do what I used to do," or, "I don't have the energy and stamina I used to have." What you are feeling is the reflection of something very basic that is going on inside your body. Starting as early as our third or fourth decades, our bodies suffer a gradual decline in their ability to produce energy. The results of this personal energy crisis are palpable in terms of how we look, feel, think, work and play—and ultimately, how we age. Let's take a look at those two scenarios again. You know how it feels on the outside, but here's what's happening inside.

When you wake up feeling full of life, the trillions of cells that make up your body are shifting into high gear, producing the fuel that powers your brain, your heart, and your muscles. With assembly line precision, your cells are effortlessly churning out the precious fuel that runs the body. At ten a.m. your body and mind are at their peak, and it shows. You shine at your client meeting. At noon your muscle cells are fully fueled. Your workout is both invigorating and effective. Late in the afternoon, as your co-workers are beginning to droop, your energy stores are still in surplus. And when your boss springs a surprise assignment on you that must be done by evening, you feel up to the challenge. Clearly, when your cells are making enough energy, you glide through your day.

When you wake up feeling drained and depleted, your cells are struggling to produce the energy needed to sustain your body and mind. They push and push and push, but to no avail. By midmorning you feel distracted and light-headed as your brain cells slump into energy deficit. By lunchtime your muscle cells are desperately struggling to pump enough fuel into your limbs as you huff and puff your way up a flight of steps. By midafternoon your head hurts and your body aches with exhaustion. Your poor overworked cells are screaming for reinforcements, but the cavalry never arrives. This dismal scenario repeats itself day after day. Like a car running on empty, you begin to sputter and stall. Weak and depleted, you catch every cold that comes your way. You lose your next promotion because you're not able to keep up with the young Turks at the office. You're getting flabby because you're too tired to get to the gym. You're depressed because life is not what it should be. You can't help but worry, "If I feel like this now, what am I going to feel like ten years from now?" If your cells could talk, they would tell you that your energy crisis will only get worse with each passing year.

Depending on your particular area of vulnerability, you will experience the energy crisis in a variety of different ways. Some of you may succumb to late-afternoon "hit the wall" exhaustion. Others may find themselves brain-tired, not thinking as clearly or as effectively as before. Still others may complain of being bone-tired, and feeling their youthful zest slipping away.

No matter how you experience it, the energy crisis discourages even the most driven of us. And no one is immune. We used to call it "getting older" or say we weren't "as young as we used to be." But it is not inevitable and, better yet, not irreversible.

The good news is, there is a cure. If you don't like the way you are feeling, you can do something about it, starting today. *Natural Energy* will show you how to reverse your energy crisis simply, safely, and effectively. Within a short time your energy levels will soar, and so will your spirits. You will feel stronger, healthier, and sharper. You will have the stamina you need to live a fuller, more active, vibrant life. Beyond feeling great, you will realize real benefits in your basic health.

End the Energy Crisis

The solution to the energy crisis is as simple as it is obvious: Revitalize the body's ability to produce energy. Sounds easy, right? But it really is the Holy Grail of medicine. It requires a totally new way of looking at the body, and this view is captivating top medical researchers the world over.

We all know that energy is essential for life. Everything we do—from breathing to thinking, from having sex to keeping our hearts beating—requires energy. To satisfy our insatiable demand for energy, our bodies function as walking power plants. Indeed, some scientists believe that the fundamental purpose of the human body is to produce enough energy to keep itself going! As we age, our energy-producing system starts to malfunction, and so do our other systems, and the impact is devastating and pervasive: our brains aren't as sharp, our hearts don't beat as efficiently, our muscles weaken. This personal burnout leaves us weaker, more tired, and more vulnerable to disease.

For years now, we've addressed this problem by focusing on

one body system at a time. We've tried, for example, to treat the circulatory system, the immune system, and the endocrine system almost as if they existed in separate test tubes. That's how doctors are taught, after all. Yet in the process, we've ignored the most important system of them all, the master system that produces the juice that runs all our machinery: the energy system. But if we overhaul the body's energy system, all our other systems will be strengthened as well. With this new surplus of energy, we may even be able to repair systems long thought broken beyond help. We can function at peak capacity again.

The Natural Energy Secret

The secret to feeling, aging, and living well is really no secret at all. Maintaining a vigorous energy system is the key, and that is what *Natural Energy* is all about. Think of it as the ultimate form of preventive medicine: by bolstering the energy system, we are fortifying the body against the breakdown and debility that allows disease to take hold.

Our body has a finite amount of energy available at one time, sort of like your house's electricity. If the system gets degraded as it ages, then it will work less efficiently. Whereas before you were able to do many things at once—just as your home could operate the air conditioner, hair dryer, and microwave simultaneously—now you find yourself struggling with just a fraction of the workload. In your house, a circuit breaker would pop, preventing damage to the system. Your body has no circuit breakers; instead, you just overload. Your body performs a grisly triage, trying to keep the most important systems operating—the basic life functions. Everything else gets starved: your brain gets foggy, your eyelids droop, you crave sugar highs, and sleep becomes far more interesting than sex. Something else happens.

Your body puts off the mundane task of routine maintenance for a time of energy surplus. And *that's* what opens the door to disease and disintegration. Anyone who has ever driven on New York City streets knows what happens when you defer routine maintenance. In just a little while, all hell breaks loose—and it's staggeringly expensive to make things right.

I don't want that to happen to you, and I think I can make sure it never does.

Natural Energy is nothing short of a revolutionary approach to the maintenance of your body. It is the first program designed to reverse the underlying cause of the energy crisis and return the energy system to its youthful vigor. Simply by taking two safe, highly effective, and over-the-counter nutritional supplements, along with making modest changes in your diet and lifestyle, it *is* possible to go "from tired to terrific" in just a few days. No matter if you are thirty and just beginning to experience the energy decline, or well into your later decades when you feel the energy crisis acutely, you can make significant improvements in how you feel and look by implementing my simple three-step program.

The *Natural Energy* program grew out of the experiences of my patients, who have achieved spectacular results by trying my easy-to-follow plan. But the science behind *Natural Energy* is more than four decades in the making.

Natural Energy is the culmination of research efforts from some of the most distinguished medical and scientific centers in the United States and abroad. Scientists have discovered that by reversing the energy crisis at the cellular level at which it occurs, we can restore and reenergize ourselves, eliminating many of the hallmarks of aging. I will be reporting on some exciting new studies that show that bolstering an "old," tired energy system can quite literally rejuvenate an aging animal so that it not only acts younger and more energetic but actually looks younger!

Scientists have seen these amazing results in their laboratories, but I have applied this new information in my practice. I have seen similar transformations among my patients that never fail to amaze. Within a few weeks—sometimes even a few days—of starting my program, people typically feel energized, stronger, and even happier. I've seen them dispense with now-unnecessary medication and regain their youthful vigor. In a word, they are rejuvenated.

The People's Doctor

I am a specialist in internal medicine with a practice of over ten thousand patients in Westchester County, a suburb of New York City. It is halfway around the world from where I was raised.

I was born during the post–World War II "baby boom" in Romania, then a Communist country. My family was affluent, which made life better for me than it was for most other people living in the Soviet bloc. My uncle, a well-respected doctor and the patriarch of my family, announced to my parents on my fifth birthday that I was going to follow in his footsteps. I never questioned his decision, and from that point on, everything I did was in preparation for a career in medicine. Although it may appear that I was pushed into medicine, I have never regretted it, not even for a moment.

When I was fifteen, my parents—thank God—left Romania to immigrate to the United States. After a year in Italy while we waited for our U.S. visas, we moved to New York City, where I graduated from New York University.

There was no question in my mind that I would go to medical school, but I had not anticipated one problem: It was 1971, and many medical schools were still telling women that they need not apply. In fact, when I went for an interview at my alma mater, NYU, where I had done exceptionally well as an undergraduate, the admissions officer told me that in good conscience he couldn't accept me because I was going to get married, have a family, and quit medicine! (He was half right: I have a wonderful family, but I never stopped practicing medicine.) I finally did get accepted to Downstate Medical Center in Brooklyn, one of the best medical schools in New York and a major teaching hospital.

As many women of my generation will tell you, we had to work harder and longer and do better than the men in our class to gain acceptance. I became a workaholic, but I loved every minute of it. After medical school, I did my residency at Kings County Hospital, the second-largest hospital in the country (only Los Angeles General is bigger), where I was attracted to the specialties of internal medicine and critical care and trauma.

Trauma medicine is a wonderful training ground for young doctors because it forces them to learn about the human body under every conceivable type of stress. The depth and focus of trauma practice is as mind-boggling as it is horrifying, as exhausting as it is exhilarating. It is perhaps the most challenging of all the medical specialties; at a moment's notice, you must be prepared to treat any kind of medical emergency—severe burns, heart attacks, gunshot wounds, bodies mangled in automobile wrecks. When I finished my residency I became director of the emergency room at Westchester Medical Center. I am very proud of my accomplishments during those years. I helped to develop the emergency medical system in Westchester County and started the first helicopter rescue program. Despite my action-packed days, I began to feel that I was missing something: getting to know my patients.

I was particularly frustrated by the fact that I was seeing patients only after they had gotten sick or injured, and I had virtually no role in keeping them well and whole. It finally dawned on me that so many of the illnesses and injuries I encountered daily in the emergency room—from heart attacks to strokes to the epidemic of hip fractures that afflict older people—were entirely preventable through appropriate intervention. I grew increasingly unhappy with patching up bodies that should not have been broken in the first place.

I wanted to redirect my medical practice so I could help people stay healthy. Eventually, my desire for more personal, one-on-one relationships with my patients led me into private practice.

I became one of the first women internists in private practice in Westchester. At first, my practice was so small I was able to accommodate my patients in a renovated, one-room garage. In time, the practice grew, and it became Irvington Medical Associates.

Private practice has taught me one thing that nobody learns in medical school, something that only a few of us have been taught by our patients: The absence of disease does not equal being well. Just because all the tests come back normal, it doesn't mean that someone is healthy. The key question is, "How do you feel?"

Many people come to me feeling so depleted that they are absolutely convinced they are sick (I'd like a dollar for every patient who has said to me, "I feel so bad I must have mono"), when in reality, they do not have a diagnosable illness. In fact, many of them do pass our conventional medical tests with flying colors.

But again, the absence of illness does not mean that you are well. To my way of thinking, *being* well means *feeling* well. "Well" means feeling energized and ready to take on whatever challenges life may throw your way, with zest and gusto; it does not merely mean normal laboratory results. Our medical system, however, sees it quite differently. We doctors are trained to practice medicine by the numbers. When a patient has a complaint, we run a battery of tests to see if we can pinpoint the cause. If nothing turns up, we assume everything is fine.

There is no better example of this than a story told to me by a patient about his seventy-five-year-old father. Every morning his father would wake up feeling so nauseous and exhausted that it took him several hours to pull himself together. After extensive medical tests, however, the father's doctor told him that his test results were excellent, and that he was the picture of health. Puzzled by this diagnosis, my patient asked, "How can my father be so healthy yet feel so bad?"

Good question. My patient's father may be the picture of health on paper, but his symptoms should not be ignored. The fact is, unless he feels well, he is not going to be able to enjoy life. In my book, he is definitely not well. Don't get me wrong: I am not suggesting that when a patient comes in with ambiguous symptoms, we should not run the appropriate tests and look for likely ailments. But I've learned from my patients that how they feel is even more important than their test results. When you don't feel your best, your body is trying to tell you something, but all too often we ignore it. Doctors ignore it and patients ignore it.

The term "doctor's doctor" refers to those physicians most respected by their peers. I think of myself as a "people's doctor." I believe my primary job is to serve my patients while my ultimate goal is to gain their respect. This is a radical departure from what most physicians have been taught. In medical school, we are taught that everything revolves around disease. We are trained almost exclusively in the diagnosis and treatment of illness—we spend our years of training immersed in disease. So it is no wonder that when someone walks into our offices, we regard every ache and pain as a potential disease. People are relegated to the role of "patients," and patients are sick. In the process, we lose sight of the seemingly obvious fact that the practice of medicine should be about the maintenance of health—health *care* in the most literal sense. Most people are born with strong, healthy bodies that will stay in peak condition if they are taught how to care for them.

Although I will use the latest medicines and techniques when I need to, my goal is to help the body to help itself. By that I mean giving the body the tools it needs to get the job done, whatever they may be. My personal approach is to use whatever works, whether it involves giving my patient nutritional guidance or the latest high-tech drug, or even prescribing vitamins and herbs. I believe that this should be the approach taken by all enlightened physicians.

Fatigue: The Number One Complaint

When I was a trauma specialist, most patients I saw were in crisis. There was no doubt that these people needed immediate and often aggressive treatment. In contrast, the overwhelming majority of patients who come to see me and other family practitioners are not actually sick in the classical sense. Many of them don't even have a fever or a sore throat or any diagnosable illness. They come in with one primary complaint: fatigue. In the overwhelming majority of cases, these people are not ill, but nevertheless lack the energy to lead the kind of lives they want and deserve. They wouldn't call themselves "well."

Although fatigue is the number one complaint that general practitioners hear from their patients, it is also the one that is most likely to be dismissed. The root cause of fatigue doesn't show up on a diagnostic test and can't be labeled as a specific illness. Since it can't be solved by conventional medicine, most doctors simply tell their patients that fatigue is as natural a consequence of growing older as gray hair and wrinkles. You simply have to learn to live with it.

This is sheer and utter nonsense. No matter what your age, you don't have to walk around feeling tired, wiped out, and over the hill. You can still feel young and vital, knowing that your best days are ahead of you.

The key to reversing the energy crisis is to treat it where it begins—in our cells.

Human energy production is a complicated process, but works essentially this way: The food we eat is broken down into smaller components in the digestive tract. Then these smaller components must be further broken down and repackaged—metabolized—into a form that can be used by the cells to produce energy.

Energy is created in our cells by microscopic structures called mitochondria. Mitochondria are known as the "powerhouses" of cells, because mitochondria produce a substance called adenosine triphosphate (ATP for short), which is, quite literally, the fuel that drives the body. Mitochondria occur in the highest concentrations in the organs of the body that work the hardest: the heart, brain, and kidneys. When we are young, our mitochondria produce enough ATP to keep us revved up and ready for action. However, as we age, so do our mitochondria. They lose their shape and structure, become hardened and calcified, producing less and less ATP. As a result, the body has less and less energy to keep going. Gradually, like a car that is running out of gas, the body slows down . . . and eventually comes to a complete stop. You've seen or experienced this time and time again. Think about it. Have you ever tried to keep up with children at a playground? They move from swings to monkey bars to slide without taking a break. For hours on end, they are in perpetual motion. No matter how hard we try, we can't keep up with them. Mitochondrial aging is the reason.

The energy shortage affects every cell and system in our body. Our heart, brain, and kidney function decline. Our cells, losing their ability to repair or replicate themselves, die. Our immune system falters, making us more vulnerable to infection and to autoimmune disorders, such as rheumatoid arthritis. Medical researchers now understand that people who suffer from Alzheimer's disease, heart failure, and Parkinson's disease—three of the most common diseases of aging and three diseases which are on the rise—also have impaired mitochondrial function. Even the less serious signs of aging, such as wrinkles and "middle-age spread," are now believed to be a direct result of a sluggish energy system.

Like an older car, our aging bodies require more maintenance to run efficiently. Our "gas mileage" decreases, leaving us laboring for energy. To make matters worse, we further sap our energy stores by eating the wrong food and living the wrong lifestyle. A peek under the hood shows that the engine—our mitochondria—is not running smoothly. We are running out of fuel.

Before we understood cellular and mitochondrial aging, we had to accept our inefficient engines as inevitable. There was no choice but to grow tired and to grow old! But now we know how to slow down and even reverse the decline in energy production so that we can maintain our mental and physical edge.

The impact of reversing the energy crisis is far-reaching and profound. If we can keep our energy system—our body's primary system—youthful and strong, every other system will stay young too. We will not only feel more energetic, but we will be better able to stave off the diseases that have been associated with "normal" aging, from heart disease to Alzheimer's disease to osteoporosis to cancer. Our bodies will regain the youthful resilience that time has taken away.

As with most cutting-edge ideas in science and medicine,

this information has been confined to medical journals accessible primarily to researchers and physicians. It can take years for the general public to learn about new concepts. I have written *Natural Energy* to share this information with you so that like my patients, you too can benefit from this knowledge.

The Three Rules for Feeling Energized Forever

Rejuvenating the body's energy system is the key to maintaining health and vitality, and it can be accomplished using *Natural Energy*'s three simple steps:

STEP #1: REPAIR

Nature has provided each of us with our own blueprint for feeling energized forever—literally within our own bodies. The miracle of *Natural Energy* centers around the Energy Pack, two nutrients which already exist in abundance in every cell—carnitine and coenzyme Q10 (Co Q10). Carnitine and Co Q10 are essential for normal energy production. They are produced by the body, but require supplementation because starting at midlife, these nutrients decline with age. The two nutrients in the Energy Pack are sold in health food stores, pharmacies, on line, and even in discount department stores. Although they work in synergy, each nutrient in the Energy Pack is unique.

CARNITINE: THE CAPSULE OF YOUTH

Carnitine is a vitamin-like nutrient present in breast milk, and so safe that it is added to infant formula. This extraordinary supplement can have a powerfully rejuvenating effect on the body and the mind. Maintaining a youthful level of carnitine throughout your life can help keep your body energized and free of many of the diseases associated with "normal" aging. Also sold as a prescription drug, carnitine is one of the few nutritional supplements that has undergone rigorous review by the Food and Drug Administration and has been approved to treat carnitine-deficiency disorders, rare diseases that primarily affect children. Carnitine is already being used successfully by innovative physicians in the U.S. and abroad for many problems ranging from heart disease to kidney failure, from chronic fatigue syndrome to infertility, and even for slowing down the progression of Alzheimer's disease. In particular, carnitine is a powerful natural mood elevator that can help relieve the toxic effect of living in a highly stressful world. Recently, carnitine has become a hot new sports supplement among the gym crowd, who say it increases stamina and endurance. The Italian soccer team swears by it! But these many uses are just secondary effects. Carnitine shines in its main mission: protecting and repairing the energy system.

COENZYME Q10: THE ENERGIZER

Co Q10 works in tandem with carnitine to enhance cellular energy production. Combined with carnitine, it boosts energy levels, keeping the body running smoothly and cleanly. Numerous rigorous scientific studies suggest that as older cells become deficient in Co Q10, this not only triggers a slowdown in energy production but increases the risk of cancer, heart disease, and Parkinson's disease. Adequate levels of Co Q10 could literally make the difference between life and death. In fact, some researchers believe that Co Q10 deficiency is responsible for the alarming rise in deaths from heart failure in the United States, a disease which, as I will explain, is literally a disease of the energy system.

If you want to stay strong and youthful, you need to take Co Q10!

. . .

Only by adding extra amounts of these key nutrients to our bodies can we return our power plants to peak efficiency. By allowing your levels of these nutrients to run down, you are leaving yourself vulnerable to the slow but steady destruction of every single system within your body. In the end, not only will you suffer needlessly, but you will be fast-forwarding the aging process.

STEP #2: RECHARGE

Retooling our energy factories is not enough. We also need to keep them well cared for and supplied with the best raw materials. *Natural Energy* includes a "mitochondria-friendly" food plan designed to increase your energy levels and preserve your mitochondria. Unfortunately, many people subsist on a diet that has just the opposite effect: it actually clogs the body's energy system, resulting in energy deficiency as well as weight gain.

As I will explain, I firmly believe that the energy crisis is primarily responsible for the "weight creep" that has become synonymous with middle age. We have become a diet-crazed country that embraces every new weight-loss fad—we have even risked our lives by taking dangerous drugs to lose weight. This approach has produced nothing but failure. The main reason why so many of us are overweight is that we are not burning enough calories, and consequently, we are storing too much fat.

Once you hit midlife, there is a noticeable change in metabolism that decreases the rate at which you burn fat to make energy. At the same time, when the energy system slows down, there is a tendency to reduce physical activity.

The Energy Pack will help stimulate the body's ability to burn fat, which will, at least indirectly, help keep off excess pounds. More directly, the entire *Natural Energy* program will give the body the energy boost it needs so that you have the stamina to maintain a healthy level of physical activity.

The solution to the fat epidemic is simple: When we are eating the right food at the right time, and taking the Energy Pack, our energy system works at optimal levels and we are able to maintain an optimal weight. The good news is, my food plan is both easy to follow and highly effective, and allows you to eat the foods you enjoy.

STEP #3: REVIVE

You cannot experience the full potential of *Natural Energy* if you neutralize each positive step with negative actions. In Step #3, I will show you how, simply by eliminating energy wasters and embracing energy boosters, you can keep your body running energy-efficient for years to come.

Your Personal Rx

Although the energy crisis is universal, it is experienced by different people in different ways. Throughout this book, I will personalize your *Natural Energy* prescription by providing advice for people with specific problems or at different stages of life.

In Chapter 2, you will learn how to identify the Seven Signs of the Energy Crisis so you can tailor the *Natural Energy* program to meet your individual needs.

The energy-boosting program prescribed in *Natural Energy* will produce both immediate and long-term benefits for anyone who follows the program. In the short term, you will feel energized, more focused, and more in control. In the long run, you will enjoy a longer, healthier life. You will look and feel fabulous!

Chapter Two

THE SEVEN SIGNS OF THE ENERGY CRISIS

I **knew something was** terribly wrong, but I didn't know what it was. I had just returned from a Caribbean vacation with a virus that left me completely depleted. Although I was officially well—all the medical tests showed that I was fine—I was still flat on my back, exhausted and depressed. What was particularly troubling was that I couldn't shake the overwhelming feeling of fatigue that stayed with me from early morning until night. For the first time in my forty-five years, I was slowing down. I was canceling appointments with patients, curtailing all social activities, and spending as much time in bed as possible. I was miserable.

This was definitely not me. It was not the Erika who was

used to working thirty-six-hour shifts in the ER without missing a beat. It was not the Erika who after answering patients' calls all night would still rise at dawn to have breakfast with her daughters, and be wide awake for her tennis match in the evening. It was not the Erika who could bounce back from a cold in twenty-four hours. It was not the Erika used to going nonstop.

I began to do research on fatigue, and found that there was little written about fatigue in healthy people. There were volumes, however, written on fatigue in illness. In fact, there were some diseases in which the primary symptom was fatigue, and those diseases often had one thing in common: They were in some way connected to the inability of the body to make enough energy.

It struck me that even though I was well, I had many of the same symptoms as people suffering from energy deficiency. As I dug deeper into the research, I became aware of a new and exciting field that linked many of the symptoms we consider normal parts of the aging process to the slowdown of the energy system that begins in midlife. My debilitating exhaustion, the inability of my immune system to easily defeat the virus, my sudden dip in mood, and the general "slowing down" that I had experienced were not only strikingly similar to symptoms suffered by people with energy deficiency, but they were also synonymous with "growing old."

It began to dawn on me that many of the complaints that I was hearing from patients were also rooted in the same problem. I wondered if some of the same techniques used to treat energy deficiency in sick people would work in basically well people who, like me, were beginning to experience the first signs of an energy decline. I also began to wonder if fatigue was not merely a symptom of many different illnesses but, ultimately, the cause.

When I began to research the causes of fatigue, I must admit that I still had the mind-set of a conventional doctor. I was looking for the cure in the form of a high-tech, expensive wonder drug that would "kill" fatigue the way penicillin kills infection. I was astonished to learn that the cure for fatigue was not to be found in a pharmaceutical house but within our bodies. The "cure" was two natural nutrients normally produced by the body—carnitine and Co Q10. These nutrients had been used in Europe and Japan for more than two decades safely and effectively, but most U.S. physicians, like myself, were unaware of them. Study after study reported that these nutrients were safe even at high doses, and caused no side effects. Frankly, I could not make the same claims for many of the prescription drugs I gave to my patients. I am a very cautious person by nature, and do not take a vitamin pill without first checking it out carefully. Confident that these supplements passed my rigorous scrutiny, I became my first test case. Within two weeks, I felt better, markedly better. Within a month, my stamina had returned. My program worked so well for me that I prescribed these supplements to other patients as part of an overall energy-enhancing program, with equally good results.

The Seven Signs of the Energy Crisis

Part of the *Natural Energy* program is teaching people how to identify the early warning signs that their energy system is beginning to decline, so that they can take steps to bolster it before the effects are too severe.

Based on my patients' experiences, and my own, I have compiled the Seven Signs of the Energy Crisis, a list of the ways in which the body tries to warn us that the energy system is burning out. Unfortunately, most of us aren't listening.

Fatigue may be the number one complaint that we physicians hear from our patients, but often, by the time we hear it, our patients are literally so depleted and exhausted they can barely drag themselves out of bed. It may not be so bad for you yet. You may just be a little run-down. Your spouse may complain that you're snappish and not as much fun as you used to be. Maybe you're just too tired to stay up to watch Letterman, or the first to poop out at the party.

Whether you're in the throes of the energy crisis or just heading that way, you're not trapped. It doesn't have to be this way. The decline in the body's energy system does not occur overnight. More often than not, it wears down gradually over a span of years—even decades—before the body is so energy-deficient that we are robbed of our health and vitality. There's still plenty of time to do something about it.

Most of us experience at least one of the seven telltale warning signs of the energy crisis long before we are severely energy-deficient, yet we typically ignore these symptoms. In retrospect, long before I was flat on my back with exhaustion, I was beginning to feel the energy crisis, but I ignored it. If you ignore the first sign, I can guarantee that over time, there will be a second sign, and eventually a third, and so on. If you allow those seven signs to accumulate, you *will* eventually get sick, just as I did. You will almost certainly be fast-forwarding the aging process. And it's quite likely that you will develop a chronic disease.

Read over the Seven Signs of the Energy Crisis and make a mental note of which ones apply to you. Chances are you will identify with several. In some cases, you may say yes to all of them.

ARE YOU:

 **Hearing your mind say "now" when your body says "later"?**

 **Waking up more tired than when you went to sleep?**

Yearning for a midafternoon nap?

Walking around in a brain fog?

Overeating because you're starved for energy?

Feeling stressed out and blue?

Fearing that your sex life is stuck in low gear?

You may have noticed that each of the seven signs is accompanied by an icon or illustration of that problem. If you said yes to a particular sign, look for its accompanying icon throughout the book. In addition to the general *Natural Energy* program, it will help you identify a section that is geared to give a special boost to people with your particular problem. This is in keeping with my philosophy that even though the energy crisis is universal, every body is unique.

Exhaustion Is Not Normal

We are often so preoccupied with the burdens of daily living that we are oblivious to our own feelings and needs. Typically, patients don't come to me at the early signs of the energy crisis; rather, they wait until they are utterly wiped out. Very often, they delay until they are completely overwhelmed and drowning in exhaustion. Instead of having just one sign, they have several, or all. In many cases, their bodies have already begun to shut down in self-defense. They are battling chronic sore throats, they complain of headaches, backaches, and colds that won't go away. I've learned from listening to my patients that many of the things we associate with adulthood—tiredness, overexertion, unmanageable stress, or feeling overwhelmed and not having time for ourselves—are early signs of the energy crisis. When I ask them when they first started feeling bad, in retrospect many of them concede that they began to notice some changes months, perhaps even years earlier, but they didn't think they were important. In many cases, people assumed that physical or even mental deterioration was simply a normal part of aging and they had to learn to live with it. In other cases, they didn't feel that these symptoms were serious enough to warrant action.

They were wrong on both counts. First, there is absolutely no reason why aging must be a one-way ticket to decline and debility. Once people begin fortifying their energy systems, they will see a dramatic improvement regardless of their age or state of health. Even very old and very sick people can benefit from my energy program. Second, simply not feeling that you are functioning at the level that you should be is good enough reason to take action.

Obviously, I am a strong advocate of prevention, and believe that the earlier you begin reversing the energy decline, the better the result. That is why I urge my patients to pick up on these early warning signs while there is still ample time to do something about them. First and foremost, I teach patients that when you don't feel quite right—when you feel tired, down, and generally "off"—your body is sending you a signal that you should not ignore. It is telling you loudly and clearly that it is time to pay attention and to reevaluate your life. In sum, it is screaming that your cells are starving and that you are not giving your body the tools it needs to repair, recharge, and revive. Ignore this warning at your own peril.

What's Your Sign?

How do you know which sign applies to you? Here is a brief description of each of the Seven Signs:

Hearing your mind say "now" when your body says "later."
When your brain says "go" but you're too tired to budge, you are experiencing the universal sign of the energy crisis. Your body's exhausted, but your head is in denial.

Waking up more tired than when you went to sleep.
Even if you're not a morning person, getting up should not be an ordeal. If your energy system is working well, you will feel rested and refreshed. If you are dragging yourself out of bed every morning, you're in an energy crisis.

Yearning for a midafternoon nap.
Do you get tireder and tireder as the day progresses? Do you avoid important meetings after lunch because you know you won't be alert? The afternoon slump is a classic sign of the energy slowdown.

Walking around in a brain fog.
Do you feel dull and out of focus? Are you having difficulty concentrating at any hour of the day? Is your memory beginning to wane? The subtle but real loss of brainpower is a sure sign that your body's running out of power.

Overeating because you're starved for energy.
You feel sluggish, maybe a little light-headed, and definitely tired. Your body is in an energy low, but you think

you're hungry. You reach for a quick sugar fix and you feel good for a few minutes, but the pattern repeats itself over and over again. Soon you're putting on extra pounds. The solution? More energy, not more food.

Feeling stressed out and blue.

Are you feeling down for no reason? Are you feeling overwhelmed and unhappy? Does the thought of doing even positive things for yourself—like going to the gym, going shopping, or even going on vacation—seem like too much effort? You need an energy boost to boost your mood!

Fearing that your sex life is stuck in low gear.

There was a time when sex took precedence over sleep for you, but now the mere thought of having sex is exhausting. Put your sex life back into high gear by recharging your energy system.

Each of the seven warning signs of the energy crisis reveals a good deal more about people than simply their state of physical or mental health. It is a veritable Rorschach test of values and priorities. In fact, I have found that most people have several signs, yet tend to report only those that have the most profound effect on their lives. A newly divorced man in his fifties who is preoccupied with keeping a girlfriend happy is absolutely mortified that he is often too tired for sex, and is not concerned that he is falling asleep in his soup over a business lunch. A college professor in her late thirties may fret more about brain fog than the fact that she is putting on weight. The common link is that these people know they are not achieving their full potential, whether it is in the classroom, the bedroom, or the boardroom. What they don't know is how the depletion of their energy system is affecting every aspect of their lives.

. . .

Above, I gave you a brief review of each of the Seven Signs so that you could quickly identify which ones apply to you. Now I will describe each in more detail so that you can gain a better understanding of why you feel the way you do, and can begin to do something about it.

Sign #1: Hearing Your Mind Say "Now" When Your Body Says "Later"

Of all the Seven Signs, #1 goes to the very heart of the energy crisis. Your brain tells you *Go go go, do do do.* You want to, and more important, you think you should. After all, you've done it before, thousands of times. Yet the moment of truth comes when it's time to swing into action and you can't move a muscle.

Sign #1 is the universal sign of the energy crisis: it happens to almost everybody at one time or another. Depending on personality and outlook, each person responds differently to the battle between mind and body. The aggressive type A personality reacts with anger and frustration. Furious that they no longer have the physical wherewithal to keep up with their high expectations, these people "should" themselves into submission. An inevitable clash of wills ensues between their bodies and their brains. Eventually, the body wins, but it's a Pyrrhic victory—both body and mind are depleted.

The more passive personality gives in too easily, and is all too ready to roll over and concede defeat. Typically, these people get depressed and distraught because they think that time has finally caught up with them, and that the downward spiral is now beginning.

The most common response, however, is one of denial. While scaling back their activities, these folks are not willing to concede that anything has changed. The "It's not happening to me" personality puts his head in the sand and hopes that ignoring the problem will make it disappear.

These reactions are perfectly normal. It *is* disturbing when your mind can no longer will your body into action. But contrary to popular belief, it's not permanent. It's not that you can't do it anymore, or that your best days are behind you—quite the opposite. The problem is that you're not giving your body the right tools to work with. It's as simple as that. People who suffer from Sign #1 typically haven't caught on to the fact that things have changed and their bodies are making new demands. In fact, they often treat their bodies like a car they're planning on trading in at the first sign of trouble, not a complicated piece of machinery that comes with a lifelong lease.

Sign #1 manifests itself in many different ways. The desire to keep performing at peak capacity is a major reason why so many of us suffer mechanical breakdowns. I take exception to the "Just do it!" philosophy that has become popular of late. Trust me, if you push hard enough, you break it or break down.

My patient Rebecca, a forty-nine-year-old nurse by day and aerobics instructor by night, learned this the hard way. Married to a thirty-five-year-old fitness trainer, Rebecca is trim and taut, and until recently, positively radiated energy. This power couple loves to surf, party, and live the good life. Rebecca's troubles began while she was training for a national aerobics competition. In addition to everything else she was doing, she got up at five a.m. every day to run ten miles before going to work. Yet no matter how hard she worked her body, she seemed to be getting weaker.

I usually saw Rebecca only once a year for her annual physical, but lately she'd been coming in almost weekly with a different ailment. First, she complained of backaches, then muscle aches, then headaches, and finally she got a cold that simply wouldn't go away. "I hate to say the word, but I'm *tired,*" Rebecca said in a low whisper.

We worked her up from head to toe: I ordered blood tests, urine tests, an MRI of her back, X rays of her hips, and an ultrasound of her uterus and ovaries. All her tests were fine. But although Rebecca was a poster child for good health on paper, she was definitely not functioning the way she could or should. The solution was simple. Rebecca needed to pay attention to what her body was trying to tell her. She was trying to maintain a schedule that would be difficult for a woman half her age, at her energy prime. It was impossible for Rebecca to keep up her active life now without priming her energy system.

Another telltale sign of your mind saying "yes" but your body saying "no way" is what I call the mystery of the incredible shrinking day. If you find yourself complaining, "There aren't enough hours in the day," and, "I can't seem to fit everything I need to do into my schedule," you may be yet another victim of this common phenomenon. My patient Irene is a hardworking homemaker who puts three kids on the school bus every morning after giving them breakfast and packing their lunches, drops her husband off at the train, and returns home to do her laundry, cleaning, and errands. At three p.m. the kids return and Irene becomes the family chauffeur and homework monitor. At six Irene gives her children dinner, and prepares another dinner for her husband. By eight it's bathtime and bedtime rituals with her kids. By nine she is flat-out exhausted. Irene used to find time during her hectic day to work out at a local gym at least three times a week and attend a weekly book discussion group, two activities that she loved. But at her last visit, she told me that she no longer had time to either exercise or read.

"I don't know where the hours go, but they go," she complained. "I want to exercise. I want to do more things for me, but there's no time for anything." Clearly, Irene has run out of steam, but she doesn't see it that way. She notices only that her day is shrinking, and with it her options. Interestingly, Irene doesn't even use the words "fatigued" or "tired"—because she goes nonstop, she assumes that she is a bundle of energy. In fact, she is so exhausted that it has altered her perception of the world. To Irene, her only problem is that she is running out of hours, not energy. Without intervention, Irene will soon begin practicing a form of functional triage—she will begin scaling back on her activities, doing only what is absolutely essential to get her through the day. Eventually, without help, she will not only be cutting the nonessentials from her life; it will soon spill over into essential activities.

If you don't acknowledge the first sign of the energy crisis—if you allow your head to push your body beyond its limits—you will run your energy factories into the ground. Within a short time, your mind will say "now," and your body will say "never"!

Sign #2: Waking Up More Tired Than When You Went to Sleep

"I can barely drag myself out of bed in the morning," is a common lament that I hear from my patients. There are two primary reasons why people wake up feeling exhausted in the morning: either they are not sleeping enough hours, or they are experiencing a poor quality of sleep. Although some of us may have more difficulty getting up in the morning than others (there *are* such things as morning people and night people), the vast majority of people, after a reasonable amount of sleep, should wake up feeling restored and refreshed. Getting out of bed should not be an ordeal.

How much sleep is enough? Some people require more sleep than others. However, on average seven to nine hours of sleep should be sufficient for most of us to awaken feeling properly rested. If you don't feel refreshed in the morning, consider it a sign that you've worn your body down and are in a state of energy deficiency.

Although lack of adequate sleep can contribute to the energy crisis, it is often a symptom of the problem and not the root cause. That's right—it is often the lack of energy that leads to sleep disturbances in the first place. The slowdown in the energy system can leave us so exhausted that we may curtail our physical activities. We drive instead of walking a few blocks. We stop going to the gym or to exercise class. We sit whenever we can. This lack of physical activity, however, can make it even more difficult to get a good night's sleep. You have undoubtedly heard someone say, "I'm tired, but it's a *good* tired." A good tired is often the tired we experience after a long, satisfying day, which usually includes some form of exercise. A bad tired is the depleted, worn-out feeling we experience after a long, frustrating day sitting at a desk or in an interminable meeting, or after a long airplane or car trip. Our minds are frazzled, our muscles are aching from underuse, and we are too keyed up to sleep.

Granted, there are people who may be suffering from a bona fide sleep disorder due to either physical or emotional causes, or a combination of both. For example, women in perimenopause or menopause often complain of insomnia or frequent night awakenings, which are related to hormonal shifts. And people who are suffering from depression typically wake up early in the morning and can't get back to sleep. Even though these sleep disturbances are not specifically due to the energy crisis, they are certainly exacerbated by it and can be helped. As I will explain later, energy depletion can make it more difficult to cope with stress, and chronic stress is another enemy of sleep. At the same time, stress is also an energy depleter—it leaves us drained and tired in a bad way. The physical and emotional upheaval it causes in our bodies can create an environment that is not conducive to a good night's sleep.

For the past twenty years, I have been preaching the importance of sleep to my patients, but they are not always receptive to my message. Many of my more macho patients argue with me that they don't need sleep. They contend that sleep is for kids and they are doing fine without it. These are the folks who are particularly hard-hit when the day of reckoning comes. They are clobbered by Sign #1—their minds may say "yes," but their bodies respond with an emphatic "I don't think so." If they're lucky, this happens when they're doing some benign and safe activity, like playing tennis or nodding out in a movie, not when they are driving on a highway or trying to impress a new client.

The irony is, some people think that sleep is a time waster, but in reality, not sleeping is a bigger time waster. If you don't sleep, you move slower and get less done. This leaves less time to do all your tasks and, ultimately, cuts into your sleep time. But becoming more energy-efficient will leave time for sleep because you're sharper, faster, and more capable. You'll have time for everything you want to do—and a good night's sleep.

Newly divorced at forty-nine after a long and unhappy marriage, Kevin is trying to make up for lost time. He works hard during the week and parties all weekend, and wonders why he feels tired. "At midnight I begin to feel drowsy," he complains. He is dating a woman twenty years his junior, who doesn't understand why he can't keep up with her. "She wants to go to a jazz club at midnight on Saturday, but I have to go to sleep. What's wrong with me?"

First, let me admit that there's nothing wrong with wanting or needing sleep. When I write about Sign #2, I highly recommend sleep! But I understand that there are nights that you may want to *Go go go,* and there are times when you can and should. Kevin, however, was beginning his nights out already tired. He had exhausted himself with a week of working long hours, not eating correctly, not getting enough sleep, and allowing his energy factories to run down. Kevin was not giving his body what it needed to keep him going.

At one time it was believed that only growing children required a good night's sleep, and that the rest of us could muddle through on as little as a few hours a night. Researchers have only recently discovered that sleep is a powerful therapeutic tool. It is nature's way of helping the body to "repair, recharge, and revive," and can greatly enhance the effectiveness of my energy program.

Human beings are diurnal creatures, which means that our bodies are programmed to run on specific day/night cycles. Hormones, also known as chemical messengers, are the software that runs the program. During the day we release hormones that make us mentally alert and physically active, but at night we produce hormones that make us wind down. Our heart rate drops and so does blood pressure. Our metabolism switches to low gear. The slowdown in body functions gives our overworked cells time to focus on their own needs; instead of having to run the body at full speed, they can catch their breath and repair damaged cells. But in the midst of an energy crisis, we don't sleep enough or well enough. It's a vicious cycle. Our bodies don't have time to repair the all-important mitochondria, which leaves us more energy-depleted than before, and so on down an ever steeper spiral.

Reinvigorating the energy system almost always results in an improved quality of sleep. In fact, this is one of the most common positive side effects of my *Natural Energy* program. If a patient has problems with sleep—as some fifty million Americans do—I offer specific strategies on how to get a better night's sleep, which I describe in Chapter 8.

Sign #3: Yearning for a Midafternoon Nap

"I can't sleep at night, but by three in the afternoon, all I want to do is sleep," complains Linda, a high-powered forty-eight-year-old attorney who has been my patient for fifteen years. Linda is the type of woman who pushes herself until she drops. But recently, she has been dropping harder and faster than ever before, a classic sign of the energy crisis. Linda's vulnerable point is in the midafternoon, when she yearns to curl up on her couch and sleep. Her fellow partners at the law firm where she works have other ideas, and often schedule afternoon meetings in which Linda is expected to be sharp and alert.

Linda is not alone in suffering from the afternoon slump: it's an almost universal sign of the energy crisis. There are several reasons why so many of us get tired in the afternoon, and as many different causes.

As I noted earlier, our bodies are designed to run on full cylinders in the morning, and to peter out as the day progresses. In the morning, we produce high levels of cortisol, a hormone which makes us alert and revs us up for action. By afternoon, levels of cortisol naturally decline, leaving us feeling a bit less charged. At the same time, levels of another hormone, insulin, begin to rise, which causes blood sugar to drop. When blood sugar drops, it makes us feel tired.

Even though we are programmed to wind down in the afternoon, the drop in energy should be a gradual fall, not a drastic plunge. Nature did not intend for us to conk out in the middle of the day—rather, the slump was a way of telling us that we need to recharge our batteries. The problem is, many of us unwittingly do things that only aggravate the afternoon slump and further sap our energy stores. The wrong food for lunch—or no lunch—can leave you feeling lethargic for the rest of the day. My patient Linda would routinely work through lunch, and then try to counteract her sagging energy by drinking several cups of black coffee. This did give her a caffeine high, but it was very short-lived. Sometimes she drank so much coffee that her heart would start to pound, which made her feel jittery and out of control.

Lack of physical activity is a major cause of the afternoon slump. There's a biochemical explanation. In order to make energy, you need to move your muscles. If you don't move your muscles, your energy production will slow down. A car you leave in the garage and never run invariably stalls when you start the motor. Interestingly, my patients who are construction workers, tennis instructors, or professional dancers, or who have equally physically demanding jobs, rarely complain of the afternoon slump.

So, if at three or four o'clock in the afternoon your desk is starting to look nice, soft, and comfortable, your energy factories are trying to tell you something!

Sign #4: Walking Around in a Brain Fog

Once the energy decline begins, most of us experience all of the Seven Signs to some degree, but are primarily concerned about those that have the greatest impact on our lives. My patient Gwen, fifty-two, is a case in point. Although she suffers from the afternoon slump, and has put on weight because she tries to recharge with candy bars, the one symptom that alarms her the most is brain fog, or the inability to think clearly. People in brain fog see the world in shades of gray. Their thinking is typically muddled and disorganized, seriously compromising their ability to think creatively or problem-solve. No wonder that Gwen, who is an architect at a prestigious firm in the Northeast, finds brain fog particularly distressing.

"I can't concentrate anymore," she complained to me. "My work has suffered. When I'm standing in front of a client, I can't always find the right words. It was never like this before."

And then she asked me the question that was keeping her up at night. "Do you think I have Alzheimer's disease?"

What Gwen is experiencing is hardly unusual for people in

their fifties, when there are "normal" age-related changes in mental function that can affect memory, concentration, and the ability to learn new tasks. In fact, these mental changes are so common that the National Institute of Mental Health has given them a name—age-associated memory impairment. Since there is no official cure, people are simply told that they have to get used to it. Well, I think that's absurd. Why on earth should we grow accustomed to the loss of that which makes us ourselves—our minds—especially when we know how to prevent it?

In that vein, I want to make one important point before I go on. Many of my patients, like Gwen, mistakenly believe that brain fog is an early sign of Alzheimer's disease, and needless to say, they are absolutely terrified. In reality, brain fog has nothing to do with Alzheimer's, or any other pathological condition. Brain fog is a sign of a common problem that can affect healthy people; Alzheimer's is a serious, degenerative disease. I do, however, believe that maintaining a strong energy system can help protect the brain from the kind of damage that may lead to Alzheimer's.

At any age, exhaustion and the underlying energy depletion can severely impair brain function. If missing even one night's sleep will leave you less sharp and focused in the morning, the effect of several nights of poor sleep are cumulative. Stress can also interfere with the mind, and trigger the production of chemicals in the body that can actually destroy cells in the memory center of the brain.

Fortunately, numerous studies as well as my own experience with patients have shown that reversing the energy decline can result in a major improvement in mental function. People find themselves mentally rejuvenated. In other words, the fog is lifted.

Sign #5: Overeating Because You're Starved for Energy

Are you gaining weight? It could be that you're overeating because you're exhausted. There's a fine line between hunger and exhaustion, and people who are out of touch with their bodies often mistake one for the other. Amazingly, I have found that for many of my patients, once they solve their personal energy crises, their weight problems vanish.

The body's response to hunger and to exhaustion is quite similar. For many of us, the responses are identical. When your brain cells are not getting enough sugar—whether it is due to hunger or to fatigue—they send an SOS to your endocrine system telling it to pump out more insulin. Insulin levels rise, using up the remaining blood sugar, and levels of cortisol—the "get up and go" hormone—also drop. If you remember from Sign #3, this is precisely how the body responds to fatigue. When you are tired, cortisol levels drop and insulin levels rise. It's no wonder that so many people think they are hungry when they are actually tired. Even if your stomach is full, the sensation of exhaustion can mimic the sensation of hunger.

Eating is also a lot more convenient than going to sleep. In most cases, if you feel overwhelmingly tired during the day, there's not a whole lot you can do about it. If you're at work, or driving your car, or caring for active children, you can't take an hour or two off for a restorative nap. So what do you do? If you're like my patient John, you open your desk drawer at work and pull out one of the many packages of Twinkies or candy bars you keep on hand for such an emergency. The sugar fix works for a while, and for a few glorious minutes John is primed with energy. But then he crashes. So what does he do? He reaches back in his desk for a candy bar . . .

John is in a vicious cycle that added twenty pounds to his weight over the past decade. That, in my opinion, is going to take years off his life and deepen his already serious energy crisis.

The "eat when you're tired" syndrome has become a way of life for many Americans. In fact, the older you are, the more likely it is that you will become or already are overweight. When it comes to "weight creep," the energy crisis plays an insidious role. It's another vicious cycle. First, the slowdown in energy production results in a slowdown in metabolism, which means that you store more fat and burn less fuel. Second, the energy crisis leaves many of us so tired at inconvenient times that we resort to eating to give ourselves a quick energy fix. The combination of a slower metabolism and more food inevitably leads to excess pounds.

Moreover, that extra candy bar could even be feeding into your exhaustion. Every time you eat, your body has to work to digest the food. Blood rushes to your gut, and enzymes have to be produced to break down the food into a form that can be utilized by your cells. Excess food is stored as fat, which also involves work. Instead of making energy, you are actually wasting energy! That is why we often feel sluggish and slow after a big meal.

When your energy system is working well, however, you will have the stamina to stay alert all day without ever resorting to a sugar fix—a fix that you'll feel lousy about afterward.

Sign #6: Feeling Stressed Out and Blue

My patient Jennifer complained that she was feeling down in the dumps. Nothing was going well in her life. She didn't like any of the men she was dating, she was unhappy at work, and furthermore, she had let her membership at the health club lapse because she found exercise boring. Jennifer shot down all of my constructive suggestions with, "It won't work," "I can't do it," and other negative comments. Clearly, Jennifer was viewing the world through a veil of hopelessness. At first glance, it may appear that Jennifer is experiencing a classic case of depression, but in reality, she's got a classic case of the energy crisis blues.

When we're exhausted, even the smallest task can seem

overwhelming. Our ability to cope with stress vanishes; problems that we could normally solve with relative ease now seem insurmountable. Often, a feeling of helplessness and sadness sets in. Interestingly, as any psychiatrist will tell you, one of the telltale warning signs of depression is chronic, debilitating fatigue. Based on experiences with my patients, I have begun to wonder whether fatigue and energy deprivation are symptoms of depression, or in some cases, the actual causes.

Unrelenting stress can also trigger depression, and here again the Energy Pack can be of great value. There is no question that when we feel strong and resilient, we are better able to handle stressful situations, but the Energy Pack does even more. As I will explain, the Energy Pack is a proven stress buster because it helps regulate hormones that are released during stress.

The energy crisis can have a devastating impact on lifestyle, leading to an even more devastating impact on mood. You don't need a Ph.D. in psychology to figure out why. When you're tired, you're not going to exercise regularly, see your friends, play sports, go to the movies, or pursue any other recreational activities that are natural mood boosters. As a result, your world shrinks smaller and smaller, while your problems grow larger and larger. It stands to reason, then, that restoring energy levels is a very effective way to elevate mood in some people.

The scientific evidence is beginning to bear this out, but my own results with patients have been positively remarkable. Nearly all of the patients who take the Energy Pack and follow other aspects of the *Natural Energy* program report an enhanced sense of well-being. In fact, within six weeks of starting on the Energy Pack, four of my patients discontinued prescription antidepressants, and have not needed them since. I'm not saying the Energy Pack is the cure for all forms of depression, but I think it will help a great many people who are as down as their energy levels.

Sign #7: Fearing that Your Sex Life Is Stuck in Low Gear

"Write me a prescription for Viagra," begged my patient Tom, forty-seven, who admitted that his sex life wasn't what it used to be. "I love my wife, we'd like to have sex more often, but it never works out. We both get home late from work. By ten p.m. one of us conks out. On the weekends, we're running with the kids. Sex never seems to happen."

I can't tell you how many patients like Tom are asking for Viagra because their sex lives are stuck in low gear, but Viagra isn't going to solve Tom's problem. By his own admission, his equipment is working fine—which is all Viagra can fix—but he's missing the energy (or the desire) to use it!

There are a great many misconceptions about the cause of sexual dysfunction. In reality, sexual problems are often caused by a combination of physical and emotional factors. In my opinion, the energy crisis tops the list.

Contrary to popular belief, sexual desire does not begin below the belt; it begins in the brain. The brain tells the body to become aroused, and the body gears up for a sexual encounter. The body and the brain have to be in sync to make it happen. If you're walking around in brain fog, feeling tired, overwhelmed, and stressed out, there is no pill in the world that is going to make you feel sexy.

Whether you are married or dating, maintaining a relationship requires both time and energy. If you are short of both, your relationship will deteriorate, and so, ultimately, will your sex life.

The best way to maintain a good sex life is to maintain your energy system. First, a strong energy system can prevent many of the physical or mental problems that can contribute to poor sexual function, such as heart disease or depression. Second, when your body is fully energized and working well—when you are at the peak of your physical and mental prowess—your sex drive will stay in high gear.

The truth is, energy is the ultimate aphrodisiac!

From the Seven Signs of the Energy Crisis, it is apparent how the breakdown in the energy system affects all aspects of your life, from your mood, to your sex life, to your eating habits, to how well you sleep. The energy system is the driving force in the body that keeps everything else humming. When the energy system begins to slow down, the consequences are as far-reaching as they are profound.

In this chapter, I've explained how energy deprivation makes you feel. In Chapter 3, I'll show you how what's happening on the outside of your body is a reflection of what's happening on the inside.

Chapter Three

THE ENERGY SYSTEM: WHY OLD FLIES CAN'T FLY

The energy crisis is not unique to human beings. I see the Seven Signs of the Energy Crisis every day in my medical practice, and have experienced a few myself, but scientists have also observed this phenomenon throughout nature. In fact, every species on earth suffers a similar fate.

Much of what we know about aging is derived from studies of animals. Scientists like to study flies in particular because, unlike humans, they have extremely short life spans—within less than a month's time, a fly can go from infancy to old age—yet they are strikingly similar in physiology to humans. Age-related changes that are subtle in humans because they occur so gradually over many decades can happen in a matter of days in the life of a fly.

A three-day-old fly can spread his wings and soar to unimagined heights, while a twenty-eight-day-old fly—an old man by fly standards—may flap his wings, but to no avail. No matter how hard he tries, he cannot propel himself up off the ground. His mind may say "now," but his body won't budge. Why? When scientists examined the wing muscles of young flies, they found that their energy system was strong and vigorous. But when they examined the wing muscles of old flies, they saw tired, worn-out muscle fibers and a failing energy system. You don't have to be a biologist to figure out that without enough fuel, flies can't fly.

We humans are not all that different from flies. As we age, our energy system begins to wear out too, resulting in an energy crisis that affects every aspect of our lives. I experienced it acutely at age forty-five when, unable to bounce back after a viral infection, I found myself in a state of chronic exhaustion. But unlike those poor flies that are trapped in the downward spiral of debility and death, I didn't have to take it lying down, and neither do you. We humans have the brains and the ability to keep ourselves from being grounded!

My patient Steve is the last person I thought would succumb to the energy crisis, but when I saw him last spring, he looked like a tired and defeated man. Steve is a forty-eight-year-old advertising executive who, despite his sedentary job, had made it a point to work out three times a week, maintaining an athletic build and youthful good looks. An avid hiker and outdoorsman, Steve had always been the picture of health and vitality. The highlight of Steve's year was always the vacation he takes with his college buddies. This summer, the group was planning to go white-water rafting on the Colorado River. Normally, Steve would be savoring the prospect of a challenging float trip with close friends, but this year, he was dreading it. Why?

"I'm not sure I can keep up with the other guys," Steve confessed. "I'm not sure I'm up for the whole trip."

Steve was too tired for his vacation. For the past few months, he'd been so drained after work that more often than not, he skipped going to the gym. Instead of playing tennis and cycling on the weekends as he used to do, all Steve did now was sleep. And his sex life? Steve said it wasn't much to talk about.

Steve's litany of problems clearly spelled out energy crisis.

He glared at me. "I suppose you're going to tell me, what should I expect at my age?"

I guarantee you, you'll never hear that from me. If you believe that a slowdown is inevitable with each passing year, and that your options will narrow exponentially with age, this book is definitely not for you.

Whatever our chronological age may be, we are entitled to our dreams and aspirations. I believe that the desire to live a full and vital life is what makes life worth living at any age. You are entitled to expect the world! But if you want your body to fulfill your expectations, you're going to have to give your body what it wants and needs.

If you want to keep on going full speed, if you want to live an exciting, fulfilling life, if you want to maintain youthful vigor and vitality forever, I'm your doctor.

As I explained to Steve, aging is a lot like playing a sport. Think of it as the ultimate physical challenge. At any age, you wouldn't dream of climbing Mount Rainier, or going scuba diving, or even playing a round of golf, without the right training and equipment, and yet many of us embark on the most challenging physical task of all, aging, without any preparation. If you want to keep on top of your game, you need to be prepared. And what's the best preparation? A fit energy system.

What your body wants and needs is energy to run on. The energy revitalization program you'll read about in the pages that follow is designed to restore your body's energy system and to keep it running efficiently for your entire lifetime.

Just ask Steve, who within a month of taking the Energy Pack and making my suggested changes in his diet and lifestyle, was physically and mentally recharged and ready to join his buddies on the float trip. I met up with him by chance at the store immediately following his vacation. He looked trimmer, stronger, and full of vitality. Steve happily reported that the trip had gone off without a hitch, and he and his buddies were already planning next year's adventure.

One of the primary reasons why my *Natural Energy* program works so fast and so well is that I do not simply write out a prescription and leave my patients to their own devices. As I see it, part of the job of a physician is to educate—that is, to increase awareness and understanding so that people know how to take care of their own bodies. I have learned that unless patients are fully informed so they are motivated to be part of the solution, they will not adhere to a treatment plan. And in order for them to fully comprehend the importance of preserving the body's primary system—the energy system—they also need to understand what is happening to their bodies as they age.

Mitochondria: The Cellular Powerhouse

The term "energy system" conjures the image of a central factory located somewhere in the body that produces energy and pumps it to wherever it is needed. In reality, there are trillions of energy factories spread out among virtually every cell of the body.

Energy production begins in our cells, or more specifically, in the tiny structures called mitochondria. Scientists speculate that hundreds of millions of years ago, long before there was anything on earth vaguely resembling life as we know it, mitochondria first appeared as primitive, single-cell organisms. The earliest mitochondria were little more than strands of DNA, the genetic "software" which runs the cell and contains the genetic code for every living thing. These early mitochondria existed for one purpose—to make energy. In fact, it is believed that mitochondria were so busy making energy that they simply did not have the time for another essential activity: reproduction. To ensure their survival, mitochondria merged with other cells that *could* reproduce, and took over the job of energy production within those cells. Today, mitochondria are found in the cells of all plants and animals, from the smallest, most primitive forms of life to highly evolved human beings.

Every organ in the body—the heart, the brain, the liver, the kidneys, and so on—is made up of hundreds of thousands of cells. Each cell is a microcosm of the entire body, performing the same tasks that the various organ systems of the body must perform. Each cell is encased in a protective membrane or "skin" which screens what gets in and out of the cell, as well as holding the individual cells together. Similar to human beings, every cell must eat, breathe, dispose of waste, and produce enough energy to keep itself going. The "brain" of the cell is the nucleus, the home of the DNA that controls every aspect of cellular function; it tells the cell what to do, and when to do it. Cells also have a digestive system called the vacuoles which control waste disposal, and a circulatory system which is the cytoplasm or watery portion inside the membrane. Supporting all this are the energy factories of the cell, the system that keeps all the other systems running—the mitochondria.

Mitochondria are found throughout the body, with a few notable exceptions including the red blood cells and the lens of the eye. Depending on the type of cell they belong in, mitochondria come in various shapes and sizes. Most cells contain from five hundred to two thousand mitochondria; therefore, every organ system contains millions of mitochondria. The more hardworking the organ, the more energy it demands, and thus the more mitochondria it has. For example, your heart, which must work continuously pumping blood throughout your body, is particularly rich in mitochondria. And your brain, also on twenty-four-hour duty, is awash in mitochondria.

Nearly all of our mitochondria—99.9 percent—are inherited through our mothers. Not all the mitochondria in your body, however, come from the same maternal ancestor. For example, the mitochondria in your heart may be your mother's mitochondria, but the mitochondria in your liver may be your great-grandmother's. Your mitochondria are the genetic link to your past, carrying genetic messages from hundreds of thousands of years back.

Mitochondria are not just the energy producers of the body; they perform numerous other important jobs. For example, mitochondria utilize cholesterol to form the raw material to make hormones, the chemical messengers which regulate all bodily functions. In the liver, mitochondria are instrumental in the urea cycle, which breaks down and eliminates waste products like ammonia from the body. And although they are not found in red blood cells, mitochondria produce the raw ingredients to make them. In sum, without mitochondria, we would not exist.

Getting Energized

Although it is but a part of their workload, making energy is arguably the most important job performed by mitochondria. The process of transforming the nutrients from food into energy is called metabolism. In a two-step process, mitochondria take materials broken down from foodstuffs (glucose, amino acids, and fatty acids) and create the fuel that runs the body. The first part of the process of making energy, the Krebs cycle, results in the production of the substance called ATP (adenosine triphosphate), a high-powered fuel. In the second part of energy production, beta oxidation, ATP is burned to power the building of molecules necessary to run our cells.

In order for the mitochondria to make energy, they need the right substrates, or raw materials. You can't make a cake without flour, and you can't make energy without the right ingredients. Two of the key substrates involved in energy production are the supplements in the Energy Pack, carnitine and Co Q10. Carnitine transports the fatty acids into the mitochondria to make energy, while Co Q10 is part of the reaction that produces energy. Without enough carnitine and Co Q10, you cannot make energy.

You've undoubtedly heard it said that you've got to spend money to make money; the same principle applies to making energy. You've also got to spend energy to make energy. When your mitochondria are working well, it costs one ATP molecule to produce three ATP molecules. However, if the energy system begins to falter or if key ingredients involved in energy production are missing, the cost of making energy could actually exceed the amount of energy produced. You could be spending one ATP molecule, but only producing two in return, or a less potent fuel, adenosine diphosphate or ADP. In other words, instead of making premium, you're forcing your body to run on a cheaper, inferior fuel. Eventually, cells that are starved for energy will die off, resulting in a gradual slowdown of every organ system in your body. Just think of the kind of damage that can cause.

The Currency of Life

Energy is the currency of life—without it, we would die. In a sense, our entire existence revolves around the production of energy. We eat, we sleep, we breathe, all to make energy. On one level, energy is simply the fuel that runs the body, much as gasoline runs a car. But that explanation is far too simple for a complex piece of machinery like the human body. Unlike a car, which has one basic job—to move people from place to place—the human body carries on thousands of different functions. All of them—including the process of making energy—require energy.

Like young flies, when we are children we make an abundance of energy, which is quickly put to good use. Childhood is a time of fantastic growth and development. Within a dozen or so years, we go from being tiny infants with underdeveloped organs to full-size young adults functioning at our peak. To accommodate these growth spurts, every second of every day we are making fresh new cells. The resilience of childhood is legendary. Children can play for hours nonstop in a playground. When children get sick, they bounce back quickly. Yet despite the tremendous workload placed on their bodies, they have energy to spare.

As we get older, and move from our third to our fourth decade, the picture is not quite as rosy. Slowly but surely we begin to show signs of energy depletion. Whereas at one time we were able to go nonstop without needing time to recharge, we now find that we tire more easily. At the same time, for many of us life is becoming ever more demanding. Many of us are simultaneously raising children (who never seem to stop), pursuing a career, running a household, and trying to maintain our bodies and minds. Our brains are telling us to "do this, finish that, go there," but our bodies are beginning to rebel. Our datebooks may be jam-packed with places to go and things to do, but the activities we once did easily and effortlessly now exact a steep toll. In sum, we become exhausted.

We experience this slowdown in terms of feeling less energetic, but the loss of energy is far more profound. In fact, it can make the difference between life and death. Think about it. When a young woman fractures a hipbone, it heals rapidly and is not considered a life-threatening condition. While it's true that the injury may cause pain and inconvenience, it is rarely serious. In no time, she's back in action. When an older woman fractures a hip, however, not only does it take much longer to heal, but she runs a twenty percent risk of dying from complications arising from the injury. Even if she survives, she stands a good chance of landing in a nursing home, or living out her days in a wheelchair.

This same scenario is replayed time and again throughout the body. The target tissues or organs may change, but the ending doesn't get any better. When you're young, the extra burden placed on your cardiovascular system from playing a lively game of tennis does no damage to your heart, but when you're older, it can bring on a heart attack. When you're young, a few extra minutes on the Stairmaster may cause nothing more serious than a charley horse, but when you're older, the same workout could blow out your knee joint. Why is it that the same stressors we can easily handle in youth can literally kill us in old age? It all goes back to energy, or more precisely, the lack of it.

We need energy to do our daily activities, but that is just one small job of the energy system. Like a master juggler, our bodies must keep several balls up in the air at the same time, or risk disaster. We need to produce enough energy to run our bodies so that we can keep our hearts beating, our brains thinking, and our muscles moving. But every time our hearts beat, we think a thought, or we move a muscle, we wear out some of our cells. We also need to produce enough energy for maintenance and repair so that when cells get injured or die, we can fix them or make new cells. For example, if we are injured, not only do we need enough energy to keep all of our body systems functioning normally, but we require an extra energy boost to heal the wound.

When we are young, we have an abundance of energy to cover all these bases, but as we get older, the decline in energy production makes it increasingly more difficult to meet these extra demands. We have a finite amount of energy that must be divided up. An added stress such as a fracture can deplete the body of the energy it normally needs to keep other systems functioning. Not only will an injury take longer to heal, but it can leave us so run down and depleted that we are vulnerable to any infection that comes our way. That is why an older body is more susceptible to disease than a younger body; the same challenges that we could overcome quickly in youth can become insurmountable in old age as our energy system winds down. Have you ever wondered why you're so tired after surgery? It's not shock. It's an energy crisis. It's a reapportioning of the body's energy system.

Common diseases associated with aging—cancer, heart disease, Alzheimer's disease, and osteoporosis—are rare in young people, but the risk of developing them rises exponentially with age. A body that is struggling to perform the basic tasks of living cannot muster up the energy to kill cancer cells, heal an ailing heart, repair damaged brain cells, or grow new bone. The fundamental difference between a young body and an old body is that a young body has the resilience to resist disease, while an old body is quickly defeated.

The fact that we become less energetic with each passing year is hardly a novel concept. The age-related slowdown in physical function must have been obvious even to our most primitive ancestors. Today we understand the root causes of the energy crisis and we also understand how dangerous it can be. The exciting news is, we know how to cure it.

The Cellular Slowdown

Long before we experience the first signs of the energy crisis—long before "I can't do what I used to do" becomes a familiar mantra—changes are occurring on the cellular level that will eventually have a profound effect on everything we do.

When it comes to aging, the energy system is the most fragile part of the body. In fact, it is so vulnerable that scientists have dubbed mitochondria the "Achilles' heel" of the cell. It's no wonder that after decades of nonstop work, these cellular wunderkind begin to show signs of burnout. As we get older, not only do we have fewer mitochondria, but our mitochondria begin to show telltale signs of aging. As we begin to see signs of gray hair and wrinkles on the outside of our bodies, our mitochondria are aging too.

To fully understand what happens to our mitochondria, and how they affect when and how we age, we need to know a bit more about them. You can't see mitochondria with the naked eye, but if you viewed one under a high-powered microscope, you would see that it looks a lot like a cell. Similar to cells, mitochondria have an outer membrane and an inner membrane which folds in on itself to create shelflike structures called cristae. Cristae are important because they provide the "table" on which energy is made—they are where the action takes place.

The raw ingredients needed to make energy are shuttled across the outer mitochondrial membrane into the inner membrane, where ATP is produced. Once the energy is produced, it must be transported back into the cell from the energy factory through the mitochondrial membrane.

As anyone who lives near a factory knows, manufacturing processes often result in an undesirable by-product: pollution.

Cellular factories are no different. In fact, every time mitochondria make energy, they also produce toxins that must be eliminated before they can do harm.

Among the primary pollutants produced during energy production are unstable forms of oxygen. Oxygen, so essential for life, is burned in the process of producing energy. But this process also creates unstable, highly reactive molecules called free radicals which damage healthy cells. These free radicals cause a great deal of harm. If they attack the DNA within the cell nucleus, they can cause changes in the cell that can lead to diseases such as cancer, heart disease, and even premature aging. The body relies on substances called antioxidants to defuse free radicals. Antioxidants include substances we produce in our bodies, such as glutathione and superoxide dismutase (SOD), as well as those we obtain through food, such as vitamins E and C. As we age, we produce less of these natural antioxidants, and as a result, there is a proliferation of troublesome free radicals. They hit us the hardest in our weak spot—our mitochondria.

The mitochondrial membranes are key to energy production. If the outer membrane doesn't function correctly, the raw ingredients will not be passed to the inner membrane. If the inner membrane becomes worn out, energy isn't produced properly, and the toxins build up. Over time, the mitochondrial membranes begin to show signs of wear and tear, losing more than half of their fat and water content, becoming rigid. A vicious cycle ensues in which the mitochondrial membrane is first injured by toxins, and then subjected to even more injury because it is unable to get rid of the toxins it is still producing. The cellular "house" gets filled with waste, which ultimately results in a slowdown in energy production, and the buildup of yet more toxins. Eventually, the entire cell will be affected, and if enough cells are damaged, this will interfere with the function of the organ.

The Missing Mitochondria

Because the mitochondria are at the center of energy production, and are thus the part of the cell that is hardest hit by free radicals, they age the fastest. Scientists have only recently documented, by looking at mitochondrial DNA, just how great a toll free radicals take on mitochondria. These tiny structures are unique in that they not only have their own cell membranes, but they also have their own DNA. This means that most cells have two kinds of DNA—the DNA in the cell nucleus, which regulates overall cellular activity, and the mitochondrial DNA, which controls mitochondrial function. There are hundreds of hereditary diseases associated with abnormal mitochondrial DNA, many of which strike children and young adults. There is also compelling evidence that over the course of a lifetime, spontaneous damage or mutations may occur to mitochondrial DNA that can also cause problems. In other words, the DNA begins to show signs of wear and tear.

More than two decades ago, scientists speculated that free radical damage to mitochondria, and the wearing down of the body's ability to make energy, may be the primary cause of aging. Denham Harman, credited with being the father of the free radical theory of aging, first hypothesized that the mitochondria were the fatal flaw in the cell that triggered the aging process. In a paper titled "The Biologic Clock: The Mitochondria?" Dr. Harman argued that the accumulation of free radical damage to mitochondria was what caused death. He asked, "Is this the cause of increased fragility of mitochondria with increasing age as well as the decrease in the number of mitochondria per cell? Are these effects mediated in part through an alteration of mitochondrial DNA functions?" This was considered to be just another theory until researchers later noticed something very peculiar about mitochondrial DNA: as we age, it becomes damaged. Scientists call this type of damage "deletions."

Several studies have reported mitochondrial DNA deletions in animal and plant cells. One of the earliest sightings of DNA damage in mitochondria was reported by Australian doctor Anthony Linnane, who spotted the phenomenon in yeast, and picking up where Dr. Harman left off, speculated that DNA deletions were related to the aging process. Since then, Dr. Linnane's laboratory has been the site of groundbreaking mitochondrial research, which I'll discuss later. It wasn't until 1992 that researchers actually observed similar DNA mutations in human cells. A well-known group of mitochondria specialists at Columbia University published a landmark paper, "Accumulations of deletions in human mitochondrial DNA during normal aging: analysis by quantitative PCR," documenting that mitochondrial DNA deletions occur in the muscle tissue of healthy older people at a much higher rate than in younger people. In fact, when researchers examined one specific mitochondrial DNA deletion in muscle cells, they found that there was a ten-thousand-fold increase during the course of the normal human life span. Not all the mitochondrial DNA in a particular muscle cell is affected by deletions, but over time, enough may be damaged to severely hamper the cell's ability to make energy. Although these researchers are cautious about making the logical leap—that is, linking mitochondrial DNA deletions to aging—they do conclude their paper by saying, "These findings have lent credence to the hypothesis that mitochondrial dysfunction is a key element in aging."

The discovery of mitochondrial DNA deletions in human muscle is so significant because the loss of muscle strength is a telltale sign of aging. In fact, starting at about age forty, we lose from two to four pounds of muscle mass every decade. The effect can be cumulative, and in some cases, muscle fibers can deteriorate to the point at which an older person has difficulty simply getting up from a chair.

Recently, these researchers discovered that not only is the mitochondrial DNA in the muscle cells of older people more likely to show signs of damage, but in some cases, it simply vanishes—that is, some old cells do not have any mitochondrial DNA at all. Although the significance of this finding is not fully understood, it is obvious that without DNA directing their activities, the mitochondria cannot make energy. In fact, they can't do much of anything.

Mitochondrial DNA damage is not unique to humans. It is the probable reason why old flies can't fly. The muscle cells of a twenty-eight-day-old fly have mitochondrial deletions that are not present in a seven-day-old fly. The muscle cells of an eighty-year-old man have mitochondrial DNA deletions that are not present in a three-year-old child. The same forces that are grounding the "old" fly are causing a slowdown in older humans.

Compared to nuclear DNA, mitochondrial DNA is much more likely to show signs of damage. In fact, depending on the site of the cell, the mitochondrial DNA can be up to ten times more likely to show signs of injury than the nuclear DNA.

Since our bodies have trillions of cells, the loss of a few cells is inconsequential. However, depending on lifestyle and genetics, some organs may be more affected by mitochondrial DNA mutations than others. Mitochondrial DNA in "hot spots" of the body, such as the brain and the heart, may be especially prone to damage. We don't think of the brain as a particularly active organ, yet our brains are veritable hotbeds of activity that devour energy. Damage to mitochondrial DNA in the brain is believed to be a factor in diseases such as Alzheimer's, Parkinson's, and stroke.

Heart failure, the inability of the heart to adequately pump blood throughout the body, is a classic example of mitochondrial breakdown. Your heart is programmed to work round the clock, beating about a hundred thousand times a day. Heart muscles consist of two million specialized cells called myocytes which are rich in mitochondria. Some cells of your body can divide and produce new fresh cells, but heart cells cannot. When they are damaged, they must be repaired or they die. A heart attack or a viral or bacterial infection can kill large numbers of heart cells. The remaining cells have to work even harder to keep the heart pumping, and over time, they may become exhausted. The heart muscle can no longer contract with enough vigor to adequately pump blood. The primary symptom of heart failure is sheer exhaustion, which is a mirror of precisely what is happening to the heart. Heart failure is in fact the result of a glitch in the heart's energy system. The mitochondria in the surviving cells are being overworked to the point at which they can no longer make enough energy to keep the heart going. Like a factory that has been overly downsized, the remaining mitochondrial "workers" become so overwhelmed that they go on strike. Not surprisingly, the incidence of mitochondrial DNA deletions is much higher in an old heart—or a sick heart—than in a young, healthy heart.

Initially, we may experience the energy slowdown in seemingly unimportant ways: nodding out in the afternoon, losing interest in sex, or bypassing the gym because we're too tired to exercise may not seem to have earth-shattering consequences. Yet if we allow our energy factories to run down, we will eventually begin the downward spiral of disease associated with the "bad" aspects of aging. Virtually every disease—from heart disease to Alzheimer's to cancer—can be distilled down to a short circuit of the energy system. As you can see, the health of your energy system is about more than just looking and feeling young. It's about life and death.

Fooling Mother Nature

Of course, the $64,000 question is: Why does all this happen? Why should our mitochondria be designed to burn out? There is no question that the human body is built for obsolescence. When it comes to children, Mother Nature is a kind, concerned parent for good reason. From an evolutionary point of view, children are of paramount importance because they are being groomed to carry on the species. As we age and are no longer prime candidates for reproduction, from nature's perspective we are little more than deadwood. Our bodies are designed to wind down, and we are expected to bow out gracefully.

Most of us, however, have other ideas. We expect to live for decades after our reproductive years are over, and we also expect to live out those years in strong, vital bodies. Maintaining a well-functioning energy system is the key to fulfilling those goals.

The whole point of *Natural Energy* is to trick Mother Nature into thinking you are young, and worthy of her time and attention. That is precisely what the Energy Pack does, which I will describe in Chapter 4.

Chapter Four

THE ENERGY PACK: **CARNITINE** AND **COENZYME Q10**

At a world-famous laboratory at the University of California at Berkeley, a team of top scientists accomplished a feat that was once considered impossible. They transformed animals that were tired and old—the human equivalent of eighty-five years old—into animals that looked and acted like youngsters. "Spectacular rejuvenation" is how one of the distinguished researchers described the effect.

At a clinic in Tyler, Texas, heart patients who were so severely ill that they were given only weeks or days to live are literally being brought back to life by cardiologist Peter Langsjoen.

At a medical research center in Melbourne, Australia, world-famous scientists have discovered how to make old, sick heart cells beat with the energy and vigor of young, healthy ones. Their discovery could revolutionize our approach to the treatment and prevention not only of heart disease, but of disease in general.

At hospitals in the United States and Italy, researchers report that they have found a treatment that appears to slow the progression of Alzheimer's disease. It is the first treatment that produces an improvement in both brain chemistry and Alzheimer's symptoms.

These scientific breakthroughs were made possible by our new understanding of the importance of the energy system, how it works, and most important, how to revitalize it.

The energy system is the body's primary system, and it fuels all other systems within the body. When we are young, our energy system is strong, but as we age, it begins to falter, and as it fails, it brings down every system with it. If we do nothing to stop its decline, it will age us before our time, stealing years from our lives.

What our energy system needs can be found in the Energy Pack. The secret is quite simple—two nutrients sold over the counter in health food stores and pharmacies: carnitine and coenzyme Q10, known as Co Q10 for short. Both are naturally occurring substances produced by the body. Carnitine and Co Q10 are instrumental in restoring and maintaining good mitochondrial health so that our energy system can function well.

Carnitine and Co Q10 are often described as vitamin-like substances because they are not vitamins in the classic sense of the word. A vitamin is usually defined as a substance that cannot be produced by the body but must be obtained through food. However, the more we learn about the functioning of the human body, the more archaic the rigid interpretation of the word "vitamin" seems. Although carnitine and Co Q10 are produced by the body, the levels we produce decline with age. It is extremely difficult to get enough of these nutrients through food alone. Therefore, we need to take supplements to boost them back to youthful concentrations. Both carnitine and Co Q10 are powerful by themselves, but when they are combined, the effect is pure magic.

Despite being largely unknown to the public, carnitine and Co Q10 have been the subject of intense scientific interest for more than four decades. There have been literally thousands of studies published on these supplements which prove them to be both extremely safe and highly effective. For more than twenty years, carnitine and Co Q10 have been used separately and in combination to treat heart disease and other ailments. In the past decade, the Energy Pack has become a favorite among competitive athletes, who take it to enhance endurance and stamina. Recently, researchers have focused on both supplements as a tonic for symptoms normally associated with aging.

Feeling Stronger, Feeling Better

In my clinical practice, I have prescribed the Energy Pack to hundreds of primarily healthy patient, ranging in age from twenty-eight to seventy-three, who were suffering from one or more signs of the energy crisis. The results have been overwhelmingly positive.

Within two to four weeks of starting their supplement regimen, patients on the Energy Pack report a remarkable improvement in energy level.

Patients on the Energy Pack notice that they fatigue less easily and have more stamina.

Those who exercise regularly report greater exercise tolerance: they are able to work out longer before tiring.

Patients report an elevation of mood and an enhanced sense of well-being. They not only feel stronger but feel happier.

"The biggest change is that I used to push myself until I crashed. Now I can do much more with less effort," said Jim, a forty-two-year-old accountant and father of two teenagers who echoes the sentiments of many of my patients. I'm not suggesting that the Energy Pack will make all your problems vanish, but it will definitely help you to maintain your equilibrium. As another patient so aptly put it, "I still have the same problems, but when I'm not exhausted, I can cope with them a little better."

UNEXPECTED BENEFITS

Boosting the body's energy system not only yielded the expected result—more energy—but also produced some unexpected benefits. In fact, four of my patients taking prescription antidepressants were able to discontinue the drugs after taking the Energy Pack for only four weeks. Experiences such as these have forced me to rethink some basic assumptions about depression. Chronic fatigue is considered a classic symptom of depression, but I now believe that in some cases it is not a symptom of depression but the actual cause. Obviously, when you are so exhausted that you cannot get through the day, you will feel upset, frustrated, and angry. The "hopeless/helpless" dynamic may be the hallmark of depression, but it is also the end result of chronic, debilitating fatigue.

Another unexpected result was that many patients noted an improved quality of sleep and for the first time in years woke up feeling well rested and refreshed. A case in point is my patient Michelle, a thirty-four-year-old mother of three children ranging in age from ten years to ten months. A self-described "walking zombie," Michelle told me that since her new baby was born, she hadn't had one good night's sleep. For the first few weeks, her baby had colic and was up every hour. Although the baby had been sleeping through the night for several months now, Michelle had not been able to return to her normal sleep patterns. When I prescribed the Energy Pack to Michelle, her eyes were glazed and her hair literally stood on end. After three weeks, she returned to my office looking and feeling wonderful. Why? Michelle was enjoying the benefits of enhanced energy and a glorious night's sleep, courtesy of the Energy Pack. Her normal sleeping patterns had been restored, and Michelle was feeling like a whole person again.

One way of testing how well a drug or supplement works is by taking patients off it for a period of reassessment. Physicians call this a "washout" period, in which we allow the substance we are testing to be eliminated from the body. After eight weeks, I routinely take patients off the Energy Pack to see whether they feel any different when they are not taking their supplements. After four weeks, I give them the option of restarting their supplements.

More than ninety percent of my patients have reported they felt so much better on the Energy Pack that they had to go back on it. Clearly, bolstering the body's energy system has far-reaching consequences on both the body and the mind.

I don't just prescribe the Energy Pack to patients—I take it myself. I have found that it has given me the added boost I need to run a busy medical practice, write my books, do my research studies, raise two daughters—and have energy to spare to wake up in the morning and do a hundred sit-ups. I am not a superwoman, but I do feel supercharged! And so can you.

How can two over-the-counter supplements make such a tremendous difference in physical and emotional well-being?

The answer to this question can be found in the abundance of scientific evidence that explains the full impact of the Energy Pack on the body and the mind.

Carnitine: The Capsule of Youth

I call carnitine the "capsule of youth." It is a true rejuvenator. We need it when we are young to grow into strong adults, and we need it when we are adults to stay young.

Carnitine is an amino acid, a building block of protein, which is produced naturally in the body and is found in food, primarily red meat and dairy products. Lamb and beef have high amounts of carnitine, but plant-based foods contain very little. Our bodies produce about twenty-five percent of the carnitine we need; the rest must be obtained through food or supplements.

In the body, carnitine is synthesized from another amino acid, lysine, along with the amino acid methionine, vitamins C, niacin, B6, and iron. If you are lacking one of these nutrients, you will not be able to make enough carnitine. In fact, many of you probably know that severe vitamin C deficiency can lead to scurvy, a disease characterized by brittle bones, destruction of connective tissue, and extreme muscle weakness. At one time, scientists believed that the muscle atrophy typical of scurvy was due to the lack of vitamin C, but today they believe it is caused by the inability of the body to produce enough carnitine in the absence of enough C.

We need an adequate supply of carnitine from our earliest days of life through our last. Fetuses get carnitine from their mothers in the womb, which causes carnitine levels to drop in women during pregnancy. Infants cannot easily produce carnitine until they are six months old, and therefore must get it from either breast milk, cow milk–based formula, or carnitine-enriched soy formula. Without carnitine, children cannot grow properly, and develop muscle weakness, heart disorders, excessive fatigue, learning disabilities, and countless other life-threatening problems. Although they are relatively rare conditions, as many as 375,000 infants are born each year with one of several medical disorders that interfere with the ability to produce carnitine. (Carnitine, sold under the brand name Carnitor, has been approved by the FDA to treat carnitine-deficiency syndromes in children.)

Neither children nor adults can survive without carnitine. It is absolutely essential for life, and it performs two vitally important jobs in the body:

The Fuel Pump: Carnitine transports fatty acids across the mitochondrial membrane into the mitochondria, where they are made into energy. In this respect, carnitine is often compared to the fuel pump in an automobile. Your body can have all the essential ingredients necessary to make energy, but without carnitine, it cannot get them into the mitochondrial "engines" where the conversion to energy takes place.

Waste Removal: Carnitine helps remove waste products from the mitochondria produced during energy production, enabling them to be eliminated from the body. If there is not enough carnitine on hand to do this job, energy production will suffer. Even worse, toxins will accumulate in the mitochondria, causing damage to its DNA, which of course, will further slow down energy production. Without adequate clean-up, toxins can also build up in other parts of the cell, damaging the DNA in the nucleus, which controls heredity and overall cell function.

Although these are the primary jobs that carnitine performs in the body, there are numerous roles it must play:

Builds Hormones: Carnitine provides the chemical backbone for the production of hormones, the chemical messengers of the body that regulate bodily functions.

Controls Fat: Carnitine helps maintain normal levels of important lipids such as cholesterol and triglycerides. Elevated lipid levels are a risk factor for heart disease and stroke.

Protect Against Blood Clots: Carnitine prevents red blood cells from clumping together to form a clot. If a clot lodges in an artery, it can block the flow of blood to the heart, which will cause a heart attack, or to the brain, which will cause a stroke.

Strengthens Cell Membranes: Strong membranes keep cells functioning at peak capacity, and protect against viruses.

Makes Red Blood Cells: Carnitine is necessary for the production of porphyrin, which is needed to make red blood cells.

Carnitine is produced in the brain, heart, and kidneys, which, not surprisingly, are the most active and mitochondria-rich organs of the body. Most of the body's carnitine is stored in muscle tissue, but it is also stored in the lens of the eye, red blood cells, and testes. (Recent studies suggest that carnitine is essential for sperm motility—that is, it helps sperm move along so that fertilization can take place. In fact, carnitine supplements have been given to men with poor sperm motility, with good results.)

Although carnitine is essential for life, its existence was unknown until 1905, when Russian and German scientists extracted a mysterious substance from the muscles of various animals. They called this new substance carnitine, from the Latin root *carn* meaning flesh or meat. They hadn't the vaguest idea of what it was, or what it did. More than forty years later, researchers discovered that carnitine was an essential growth factor for nonhuman organisms such as mealworms. This led scientists to speculate that humans might need it as well. In the 1950s, researchers learned that carnitine was essential for energy production, but still didn't have the whole story.

The importance of carnitine in the human body was not fully understood until the 1970s, when the first carnitine-deficiency diseases were diagnosed. The fact that severe carnitine deficiency could produce life-threatening symptoms such as heart failure and extreme muscle wasting forced researchers to take a closer look at this amino acid.

Carnitine levels naturally decline with age, and this results in less than optimal amounts in the body. This reduction in carnitine levels will not cause an obvious diagnosable deficiency disease like scurvy, but it will result in our bodies not being able to work as well or efficiently. Starved for energy, cells in key organs will begin to wear down and finally wear out. Over time, the effects are lethal. Many researchers believe that subtle carnitine deficiencies may be responsible for a whole range of illnesses, from heart disease to Alzheimer's to cancer. (Interestingly, obese people have lower than normal levels of carnitine, although they would not be diagnosed as deficient.) In fact, some researchers speculate that carnitine deficiency is a leading cause of the ultimate disease—aging.

To compound the problem, lifestyle and environmental factors may also be causing carnitine levels to drop to alarmingly low rates.

Dietary Deficiency: Our modern diet lacks carnitine-rich foods such as red meat and dairy products because they are high in unhealthy saturated fat. Yet by cutting out these foods we are also in effect eliminating our main sources of carnitine.

Sedentary Lifestyle: Our sedentary lifestyle is contributing to the decline in carnitine production. Exercise boosts the production of carnitine, but study after study documents that most people do not get enough exercise. We cannot be making enough carnitine.

Toxic Environment: Pollution may also play a role in promoting carnitine deficiency. Carnitine is key to keeping the body free of many different types of toxins. As the environment becomes more polluted, and our bodies are put in toxic overload, we place an even greater demand on our carnitine stores. The more pollution we encounter, the more carnitine we use.

The importance of carnitine cannot be overstated. Without it, life wouldn't exist. It is the life force that drives everything from the most primitive single-cell organism to the trillions of cells that make up the human body.

A TALE OF REJUVENATION

Without adequate levels of carnitine, fatty acids cannot get into the mitochondria. Without those raw materials, energy production will stall. It's as simple as that.

The decline in carnitine is a major cause of the burnout of the energy system. When our energy production stalls, it damages the machinery of our mitochondria, much as running a car without oil tears up the engine. Boosting carnitine back to optimal levels through supplementation will help to prevent or even reverse this destructive process. There is also compelling scientific evidence that restoring the energy system will have a restorative and protective effect on every other system in the body.

I have seen these positive effects of carnitine supplementation in hundreds of patients who look and feel reinvigorated within weeks of starting on the Energy Pack. But anecdotal research is one thing. Scientists have *documented* the amazing effects of carnitine supplementation in their laboratories.

Some of the most exciting work on carnitine is being performed in the laboratory of Dr. Bruce N. Ames at the University of California at Berkeley. Dr. Ames is famous for devising the Ames test, the gold standard for rating the cancer-causing potential of hundreds of common chemical substances found in food or used in the environment. The work being undertaken in his laboratory on carnitine is nothing short of spectacular.

Until recently, it has been virtually impossible to assess the full damage of mitochondrial aging to the function of particular organs. The problem was, old mitochondria are so fragile that they were often destroyed by the standard scientific procedures used to isolate them. New technology has made it possible for scientists to get a more accurate view of the full impact of the wearing down of mitochondria. Scientists at the Ames laboratory have studied the effect of mitochondrial aging on the liver function of old animals. The liver is a critical organ for our survival because it is responsible for the metabolism of drugs and toxins which are ingested through food, produced through normal metabolism, or found in the environment. The liver is also involved in the production of bile, which is necessary for the breakdown of fat. The liver is also the organ responsible for storing glycogen, used to fuel muscles, and fat-soluble vitamins. A well-functioning liver is essential for our survival.

According to Dr. Tory Hagen, formerly of the Ames laboratory and now a principal investigator at the prestigious Linus Pauling Institute at Oregon State University, mitochondrial aging takes a severe toll on the liver cells of aged animals. In fact, two thirds of the cells isolated from older livers had mitochondria that were dysfunctional. In particular, the mitochondrial membranes were damaged to the point at which their ability to produce energy was severely compromised. "You have a loss in the ability of mitochondria in the majority of cells to respond to an increased energy demand or to simply meet normal cellular energy demands," explained Dr. Hagen. "We're seeing a loss in overall energy consumption in the majority of these cells so that the overall basal metabolism is slowing down. It's getting to what people tell you clinically. 'Well, I no longer have this kind of get-up-and-go, I can't do what I used to do.' We think that there is a reason. We think it's because the mitochondria are slowing down."

What happens if you supplement older animals with carnitine? Will it help reverse the mitochondrial slowdown? In a groundbreaking study, the researchers at the Ames laboratory put acetyl-l-carnitine (a form of carnitine) in the drinking water of rats twenty-four months old, which is the human equivalent of about eighty-five years. Like older human beings, these older rats showed the telltale signs of aging: their activity level had declined dramatically, and their fur was thinner and more ragged. They looked and acted old. Within a month of starting carnitine supplementation, however, the researchers noticed some absolutely incredible changes occurring in the supplemented rats both on the inside and the outside.

Most noticeably, there was a sudden increase in the activity of the older rats. In fact, a computer camera which documented the movements of the older rats found virtually no difference between the activity level of the carnitine-supplemented rats and that of rats half their age. As Dr. Ames jokes, the rats had become so frisky "they were up all night doing the Macarena."

The carnitine-supplemented rats not only acted younger, they also looked younger. Their bodies were more muscular and their fur was shinier. It was even becoming difficult to tell the difference between the old animals and the young animals. Dr. Hagen recalls receiving a frantic phone call one weekend from the animal care people at the university, who told him that the older rats had simply vanished. Dr. Hagen rushed down to the lab to check on his experiment, fearful that someone had taken his test subjects. Much to his relief, he found that the old rats were still there. They looked and acted so much younger that the animal care people had mistaken them for the younger animals.

It also appears that the older rats were *thinking* like much younger animals. Similar to humans, as rats age they experience a decline in mental function, notably in their ability to reason and problem-solve. An older rat will have a more difficult time negotiating its way out of a maze than a younger rat. But older rats supplemented with carnitine showed a marked improvement in their ability to get through a maze, undoubtedly due to carnitine's beneficial effect on brain function. Since the brain is one of the most energy-dependent organs in the body, it stands to reason that boosting energy will boost the brain. Experimental evidence is beginning to bear this out.

This is the first study to make a connection between the rejuvenation of mitochondria on the inside of the body and the signs of rejuvenation on the outside of the body.

What precisely was causing this amazing transformation from "old" to "young"? An examination of the mitochondria of the liver cells of the newly rejuvenated older rats revealed an even more dramatic story of renewal: the mitochondria looked like young mitochondria. The mitochondrial membrane, the site at which energy is made, has been vastly improved. As a result, the great majority of the liver cells were now able to produce energy efficiently. They were humming along as well as cells isolated from younger animals. You may remember that two-thirds of the old liver cells were previously found to be mitochondrially dysfunctional. The change is dramatic, I'm sure you'll agree.

There is one downside to the experiment: carnitine slightly increased the level of oxidative stress in the cells, making them more vulnerable to free radical attack. In fact, any time you turn up the metabolic rate—including when you exercise vigorously—you are utilizing more oxygen and therefore increasing exposure to free radicals. This is precisely why I combine carnitine with Co Q10, a potent antioxidant that reduces free radical damage. Even more important, it is essential to eat a diet rich in natural antioxidants (such as vitamins C, E, and bioflavonoids), which are found in fruits, grains, vegetables, and vegetable oils.

Although the researchers did not examine the mitochondria from other organs, such as the brain and heart, presumably these showed similar benefits from carnitine. In fact, studies conducted in other laboratories have confirmed that carnitine supplementation can successfully rejuvenate mitochondria taken from the heart cells of older animals so that they function as well as young mitochondria.

It stands to reason that since carnitine supplementation can reverse the age-related decline in energy production in animals, it should have the same effect in humans. In fact, there are reams of human studies that confirm exactly that. Carnitine supplementation has been proven to fix energy systems broken by carnitine-deficiency diseases. People with heart and kidney disease who have lower than normal levels of carnitine typically show great improvement after receiving carnitine supplements.

In a sense, aging itself may be a type of carnitine-deficiency disease that can be cured by boosting carnitine levels. I'm not saying that we won't grow old, and that our bodies won't change through the years, but there is no need to grow weaker and sicker in the process. To my way of thinking, true preventive medicine is preventing the kinds of destructive changes within the body that interfere with the quality of life. We should all be doing the Macarena well into old age, and with carnitine, it's now possible!

NATURE'S BRAIN BOOSTER

When you wake up in the morning feeling full of energy—when you can get through the day free of fatigue—it will undoubtedly have a wonderful effect on your mood. However, there are other reasons why people on the Energy Pack report feeling happier. One of the primary reasons is that carnitine is a natural stress buster. It helps our bodies better cope with stressful situations that can not only wreak havoc on our mood but destroy our health as well.

Before you can understand why carnitine makes you feel so good, let me first explain why stress can be so bad for you.

Stress is the way your body responds to any situation that taxes you mentally or physically. Many situations can trigger stress, ranging from a fight with your spouse to battling a viral infection, from running for a bus when you're late for work to dealing with a difficult coworker.

Regardless of the kind of stress we are under, our bodies respond in the same way. The autonomic or involuntary nervous system takes over—the same system that regulates such involuntary activities as the beating of our hearts. The stress response is a prehistoric mechanism designed to ensure the survival of the earliest humans. The autonomic nervous system puts us in "fight or flight" mode, which means our bodies are gearing up for action. The brain alerts the adrenal glands (located on top of the kidneys) to start pumping out the stress hormones epinephrine, norepinephrine, and corticosteroids. These hormones prepare the body for immediate physical activity. In Chapter 2, I talked about how cortisol is the hormone that makes us alert, and why cortisol levels are higher in the morning when we are beginning the day than in the evening when we are winding down. But too much of a good thing can be bad. When we are under stress, we squeeze every last drop of cortisol out of our systems, which makes us hyperalert at any hour of the day or night. High levels of cortisol force blood sugar levels to rise sharply to provide fuel to burn. Blood pressure soars, blood rushes to our limbs in preparation for flight, and we are geared up for action.

This prehistoric stress response allowed our ancestors to flee an attacker or a wild animal. It was vital to human survival. Today, for most of us, stress is a very different experience, one with no physical outlet. We do not burn off stress hormones with a sudden burst of activity, as nature intended. They stay in our bodies as we sit behind our desks seething in anger, or sit home in front of the TV, sullen and hurt, or lie in bed at night worrying about the events of the day. At any age, excessive stress can take a steep toll on our physical and mental well-being, but the older we are, the worse it is.

Younger bodies are able to dispose of stress hormones and return to normal quickly. As we age, however, it becomes difficult for our bodies to turn off the stress response. Levels of stress hormones stay elevated for longer periods of time. Over time, high levels of cortisol can damage vital body organs. The aforementioned continuous rise in blood sugar increases the risk of diabetes, which in turn increases the odds of developing heart disease. Cortisol inflicts a double whammy on the heart: it can damage both the heart muscle and the arteries delivering blood to the body. It can even increase the risk of osteoporosis by blocking the growth of special cells that grow new bone. There is no doubt that chronic exposure to high levels of cortisol can accelerate aging.

Cortisol has a particularly devastating effect on the brain. It can injure brain cells, especially those in the hippocampus, the portion of the brain that controls short-term memory. Interestingly, short-term memory is one of the first signs of age-related memory loss (the so-called "normal" memory loss that begins during midlife). Some researchers believe that lifetime exposure to high levels of cortisol could be the primary cause of short-term memory loss and may even be a causal factor in Alzheimer's disease.

Why do stress hormones stay elevated in older people? The answer is simple: The brain doesn't shut them off in a timely fashion. When the adrenals pump out cortisol, the brain relies on special receptors to monitor its level. When cortisol levels become too high, the receptors alert us that it's time to shut down the adrenals. As we get older, however, the receptors become dulled, and they fail to monitor the level of cortisol as carefully as they should. As a result, the brain is bathed in high levels of cortisol for longer and longer periods of time. This not only kills brain cells but also interferes with the production of neurotransmitters, the chemicals which allow brain cells to communicate with each other. When communication between brain cells is hampered, we cannot think as clearly or function as well. In other words, we are walking around in a brain fog.

The destruction of brain cells also has a profound effect on mood. High levels of stress hormones dampen the production of endorphins, the body's natural pain busters and mood boosters. Over time, stress hormones destroy our natural coping mechanisms. Without a brain boost, stress becomes even more lethal to the body and the mind.

The good news is, carnitine can help prevent and even reverse the devastating effects of stress on the brain, while enhancing the mind's basic functioning in several important ways:

Regenerates Cortisol Receptors: Carnitine can restore cortisol receptors in the brain so that they are able to perform the job that nature intended—that is, shut down production of cortisol when it is no longer needed.

Enhances Neurotransmitters: As we age, levels of neurotransmitters decline, which contributes to so-called "brain aging." Carnitine is a precursor to acetylcholine, an important neurotransmitter that helps maintain normal brain function. In addition, carnitine helps to normalize the level of "feel good" endorphins, which enhances our ability to cope with stress.

Boosts Cellular Detoxifiers: The brain is a major energy-producing and -consuming organ. The more energy that is produced, the more toxins that accumulate in the cells. The more toxins that accumulate, the more likely it is that they will interfere with normal brain function and further hamper the production of neurotransmitters. This just further erodes our ability to cope with stress. Carnitine not only helps to eliminate toxins from the mitochondria but also to boost levels of other important natural detoxifiers, including glutathione and Co Q10.

Numerous psychological tests have documented carnitine's positive effect on mental function in patients being treated for a wide variety of psychological disorders, and I have seen it among my patients who take the Energy Pack. They not only are better able to handle the stress of daily life, but often report feeling calmer and more focused.

The combination of physical and emotional stress can be damaging to even the strongest and most resilient people at any age. For example, female athletes, who are under intense physical and emotional stress, often suffer from amenorrhea, the loss of their menstrual cycle. It is not uncommon for young women during stressful times, such as college exams, to experience a bout of amenorrhea. We shouldn't be surprised that stress would cause menstrual disorders: the menstrual cycle is regulated by a delicate interplay between hormones, an interplay easily disrupted by excess cortisol. A recent study performed in Italy showed that within a week of taking carnitine supplements, young female endurance athletes with amenorrhea regained normal menstrual function.

The carnitine effect is particularly strong in older people with severe dementia. Several studies have shown improvement in mental function in people with early Alzheimer's disease and related problems. In particular, carnitine has been shown to slow down the progression of Alzheimer's disease. While patients taking carnitine showed modest improvement in terms of verbal fluency and short-term memory, they showed considerable improvement in mood. Since there are no effective treatments for Alzheimer's, and no cure as of yet, any substance that can delay the progression of this disease or minimize its impact is worthy of more research.

Study after study has led to the inescapable conclusion that carnitine makes people mentally sharper, happier, and better able to cope. In one sense, it restores hope. The loss of hope—the feelings of hopelessness and helplessness—not only has an adverse effect on our minds, but can have a destructive effect on our bodies. Under bad enough circumstances, it can kill us.

In one experiment, scientists placed mice in a maze from which there was no escape route. Periodically, the mice received an electric shock, not strong enough to kill or even hurt them, but certainly enough to disturb them. Within a matter of days, the mice died. Why? Mice can't speak for themselves, but the researchers assumed that the mice found the situation to be so stressful and so hopeless that they simply couldn't tolerate it, emotionally or physically. It wasn't the electric shock that killed them—as unpleasant as that must have been—but rather the feelings of utter despair brought about as a result of their predicament.

Scientists repeated the experiment, but this time pretreated the mice with carnitine. Despite the escape-proof maze and the intermittent electric shocks, all of the mice survived healthy and intact. Once again, since mice can't be interviewed, the researchers speculated that the carnitine rescued the mice by fortifying their bodies against the lethal effects of stress. Instead of succumbing to the unrelenting pressure, the mice took it all in stride. We can assume that even though they were still under enormous stress, the mice were better able to cope with the difficult situation.

When it comes to stress, we humans are not all that different from mice. When we feel trapped, we are vulnerable to anger and depression, which increases the odds of developing many seemingly unrelated illnesses, from cancer to heart disease to even osteoporosis. Keeping stress under control is one important way to maintain mental and physical health. Carnitine supplementation can help provide a buffer between our bodies and stress. Although we can't always control stressful situations, we can control the impact of stress on our bodies. Carnitine alone can't do the job. In Chapter 8—"Revive!"—I give more tips on how small changes in lifestyle can yield big changes in our ability to adapt to stress.

Until now, I've focused on the effect of carnitine on the brain, but in fact, it has a profound effect on every vital system of the body.

BOOSTING IMMUNE FUNCTION

A well-functioning energy system is important for the basic functioning of the body, but it is especially critical for the maintenance of good health. As the energy system declines, so does the body's ability to fight disease. In Chapter 3, I described why a hip fracture that is an inconvenience at worst for a young person can be lethal for an older one. The underlying problem is that an older body no longer has the same stamina and endurance as a younger body. Because of our diminished capacity, so much of our energy must go to simply running the basic body systems that we have little reserve to meet additional challenges, like healing a broken bone or fighting a virus. When the energy system breaks down, it takes the immune system with it.

The immune system is not confined to one organ; rather, it is a collection of specialized cells throughout the body that forms our defense against disease. It is the job of the immune system to seek out and destroy organisms that do not belong in the body, and to weed out potential problems, such as mutated cells that if allowed to grow, may become cancerous.

The primary cells of the immune system are white blood cells called lymphocytes, of which there are several different kinds, each assigned to a particular task. T cells, produced by the thymus gland, wage battle against bacteria, fungal infections, and cancer cells. B cells produce proteins called antibodies or immunoglobulins, which hunt down foreign invaders and attach themselves onto them. Another type of immune cell, the natural killer cell (NK cell), monitors the body for potential cancers and, as their name implies, kills them before they can multiply. Still other types of cells, macrophages, gobble up foreign material found in the body. All of these different cells work together to maintain health.

As we age, our immune system ages too. This phenomenon is called immunosenescence. The immune system no longer works as efficiently or as well, for several reasons. For one thing, as our bodies slow down, we lose our ability to react quickly to the challenges of everyday living. By the time the immune system detects a bacterial or viral invader and then produces the appropriate response, the infection may have already taken hold. In some cases, the immune system loses its ability to distinguish between the body's own proteins and foreign proteins. Exposure to stress also plays a role in dampening immune function. Numerous studies have shown that long exposure to cortisol can interfere with the activity of disease-fighting T cells, as well as with the body's ability to produce antibodies.

As a result, we are vulnerable to many different types of illnesses. For example, during the winter months, older people are often advised to get flu shots because they have a much tougher time fighting the influenza virus. A virus that can be shaken easily by a young immune system can be deadly to an old one. It is no coincidence that the rate of cancer rises exponentially with age: the NK cells simply don't have the zip they once had. Older people are especially vulnerable to rheumatoid arthritis and other autoimmune diseases, in which the cells of the immune system inexplicably turn on the body's own tissues.

It may surprise you to learn that a younger body that is under extreme stress due either to exertion or to illness shows many of the symptoms of immunosenescence seen in older people. For example, endurance athletes who run their bodies ragged while training for an event are often highly vulnerable to respiratory infections due to a dampened immune response. The AIDS virus, which often affects young people, is characterized by a decline in T cell activity similar to what is seen in much older people.

Many of my patients on the Energy Pack report that they do not get as many colds and flus as they did before taking carnitine. This is undoubtedly due in large part to carnitine's beneficial effects on the immune system. Part of carnitine's positive effect is a result of its ability to control cortisol, the stress hormone which can inhibit immune response. Being ill itself is stressful and raises cortisol levels, which is a good reason for people who are chronically ill to take extra carnitine. Physicians may be reluctant to prescribe antidepressants for the chronically ill because many antidepressants can affect immune function. Carnitine is a safe way to control the stress associated with chronic illness and boost mood without inhibiting the body's ability to fight disease.

THE HEART REJUVENATOR

When I talk about the diseases of aging, there is no better example than heart disease. Unless it is due to a congenital problem, children rarely if ever suffer from heart disease. You simply do not hear about an otherwise healthy child or teenager keeling over on the tennis court or in the schoolyard from a heart attack. Such occurrences are so rare that they are reported on the news, and are often caused by a genetic or congenital problem. This is not true for adults; from midlife on, heart disease is very common in both men and women. In fact, heart disease is the leading cause of death in the Western world.

Why is heart disease rare in children, yet common in adults? The probable cause is mitochondrial aging.

Most people think of the heart as a single pump that circulates blood throughout the body. In reality, the heart is a collection of millions of specialized cells which make up the heart muscle. Each time your heart beats, it is delivering blood and vital nutrients to the organs and tissues of the body. Your heart muscle is programmed to work automatically, and even when it is supposedly resting—when you are sleeping—it is still working. Over the course of an average lifetime, the heart beats 2.5 billion times. None of those beats could happen without energy.

As one of the hardest-working organs of the body, the heart is rich in energy-producing mitochondria. Not surprisingly, there is also a high concentration of carnitine in heart cells.

Oxygen is also essential for the production of energy, which is why we can't live for more than a few minutes without taking a breath. The most common form of heart disease—coronary artery disease—is in reality an oxygen-deficiency disease. What happens is that one or more of the arteries delivering blood, nutrients, and oxygen to the heart becomes blocked. Without enough oxygen, the heart becomes energy-starved and cells begin to die. Since we are born with all the heart cells we will ever have, the loss of a substantial number of heart cells can be catastrophic.

A heart attack results when the heart does not get enough oxygen for a significant period of time, which causes damage to portions of the heart. Angina or an ischemic attack occurs when the heart is deprived of oxygen for a short time. This loss of oxygen interferes with energy production, but carnitine helps compensate by improving energy production even in these times of oxygen deprivation.

The slowdown in the body's energy system is a key factor in heart disease. When the mitochondria in the heart do not function efficiently, not only is less energy produced, but more toxins are allowed to accumulate in the heart cells. Over time, these toxins poison the cells, leading to even less mitochondrial capacity. Your heart may be barely able to accommodate the needs of daily living, and break down under the pressure of added physical or emotional stress.

This raises the question: Which comes first, the mitochondrial slowdown, or the blockages in the arteries that result in the mitochondria not getting enough oxygen to make energy? The answer is, your heart is probably taking a beating from the inside *and* the outside.

Several studies have documented that diseased hearts of any age have greater numbers of mitochondrial abnormalities than healthy hearts. In other words, the sicker the heart, the sicker the mitochondria. As we age, the number of mitochondrial abnormalities also increases within our heart cells, regardless of whether our hearts are healthy or sick. The wearing down of the mitochondria in the heart lessens heart function.

When mitochondria don't make energy, what happens to the free fatty acids that are meant to be burned for the production of ATP, or fuel? They begin to accumulate in the cells, and eventually can lead to an excess of fat (lipids) in the blood. High levels of blood lipids can form clots and blockages in the arteries, provoking a heart attack.

There is good evidence that the downward spiral of heart aging can be stopped—and maybe even reversed—by taking the Energy Pack. People with heart disease have significantly lower levels of carnitine than normal. For more than a decade, physicians at hospitals in Italy, Sweden, and Japan have given carnitine to patients suffering from angina and heart disease, with good results. Often included with other therapies, carnitine has been shown to improve overall heart function, and in particular, to improve energy metabolism in the heart.

In two landmark studies published in the prestigious *Journal of the American College of Cardiology* in August 1995, physicians in Italy and the Netherlands concluded that patients who were treated with carnitine early in the course of acute heart attacks and whose treatment was continued for twelve months thereafter reduced the incidence of ballooning out of the heart muscle, which is a common occurrence among heart patients, and can lead to heart failure. Another important study conducted in the Philippines found that patients treated with carnitine for eight weeks after an acute heart attack had smaller areas of "dead" heart muscle after the heart attack than patients not treated with carnitine. As far back as 1991, clinical researchers at medical schools throughout Italy found that patients with stable angina who were treated with carnitine for six-month periods exhibited a significant reduction in heartbeat irregularity as well as marked increase in exercise tolerance when compared to their peers who did not take carnitine.

More recently, researchers have begun to investigate whether carnitine can prevent or even reverse the kind of changes in mitochondria that weaken the heart, leaving it vulnerable to heart attack. The results are very exciting. A group of researchers at the University of Bari, in Italy, injected rats with 300 mg/kg of body weight of acetyl-1-carnitine. Shortly thereafter, mitochondria isolated from their hearts were analyzed. The researchers reported that many of the telltale signs of aging were absent in the mitochondria from the hearts of rats that had been pretreated with carnitine. In particular, they reported that an enzyme required for energy production, cytochrome C oxidase, which is thirty percent lower in older animals than in younger animals, was boosted to youthful levels after carnitine supplementation. The scientists also found that carnitine produced positive changes in the mitochondrial membrane, which is essential for energy production. In particular, carnitine restored levels of cardiolipin, an important lipid which helps maintain the integrity of the mitochondrial membrane. These changes in mitochondria were not merely cosmetic—the researchers reported that all of them added up to improved mitochondrial function. In other words, this study indicates that carnitine supplementation may help slow down heart aging.

TOO LITTLE CARNITINE CAN BE HAZARDOUS TO YOUR HEALTH

If carnitine supplementation can make old heart mitochondria young again, is the age-related decline in carnitine a contributing factor to heart aging? There is tantalizing new evidence that this is so. Dr. Dennis J. Paulson of Midwestern University gave healthy young male rats a drug that produced mild carnitine deficiency—the kind that is not serious enough to be classified a medical problem, but could over time create one. After about twelve weeks of treatment, Dr. Paulson recorded a reduction by fifty to sixty percent in the carnitine content of heart cells, but not enough to make a difference in the ability of the heart to function. Dr. Paulson then treated rats with the same drug for a period of twenty-four to twenty-six weeks. The rats' hearts performed well when given a normal workload. But when given additional work, the hearts of the carnitine-deprived animals showed significant changes in mechanical function and energy production. Over time, such changes could prove to be deadly. As with the carnitine-deprived rats, your heart may function well under normal circumstances but give out when it is suddenly confronted with a physical challenge, such as running for a bus or chasing after a tennis ball.

There is a great deal of circumstantial evidence to suggest that long-term exposure to carnitine deficiency—the kind that we experience from midlife on—could accelerate heart aging. From these animal studies, and the good results experienced by patients already taking carnitine for heart disease, we can surmise that carnitine supplements can turn back the clock on heart aging. (As I will describe later, carnitine and Co Q10 combined provide even more powerful protection against heart disease than either one taken separately.)

Carnitine supplements can also help control a risk factor for heart disease: elevated levels of the lipid triglycerides. Normal triglycerides should be under 200mg/dl; higher than normal triglycerides are a risk factor for heart disease in women, and to a lesser extent in men. Numerous studies have shown that carnitine can significantly reduce elevated triglycerides. At the same time, carnitine can also raise levels of HDL ("good" cholesterol), which moves excess cholesterol out of the body.

Thousands of scientific studies and the positive experiences of my own patients prove without a doubt that carnitine is a tonic for wellness. It rejuvenates the body and mind, and strengthens every system in the process. In the laboratory, scientists have proven that carnitine can make old animals look and act young again. In clinical studies, carnitine has been shown to be an effective treatment for a wide variety of ailments ranging from heart disease to Alzheimer's to depression to menstrual disorders. In my practice, I have seen the wonderful effects of carnitine on basically healthy people who need to counteract the energy crisis. I know that carnitine will help my patients maintain the healthy bodies that most of us are born with.

Coenzyme Q10: The Energizing Antioxidant

Coenzyme Q10—known as Co Q10—is the perfect complement to carnitine. Nature intended for these two supplements to work together. They are both critical to the production of energy in the body, and if you are lacking in either one, you will not be able to produce enough energy to maintain optimal health.

Like carnitine, Co Q10 is produced by the body and is also found in food. It is present in small amounts in many different foods, but the highest concentrations can be found in organ meats. However, in order to meet my average recommended dose of 90mg of Co Q10, you would have to eat more than six pounds of beef, fourteen pounds of peanuts, and six pounds of sardines every day! As with carnitine, Co Q10–rich foods tend to be high in fat and are rapidly disappearing from the American diet. Taking a supplement is the only way to ensure that you are getting enough Co Q10.

In the body, Co Q10 is synthesized from the amino acid tyrosine, with the help of at least eight vitamins and several trace minerals. You must have ample amounts of these vitamins and minerals to make the quantity of Co Q10 you need. Since much of the food we eat has been processed to the point at which it no longer contains adequate levels of nutrients, many of us have insufficient levels of the vitamins that are key to producing Co Q10.

Co Q10 is a remarkable molecule. Unlike carnitine, which is produced only in the liver, kidneys, and brain, Co Q10 is produced by every cell of the body. It is so ubiquitous in the body that it is also called ubiquinone.

While an enzyme is a protein found in living cells that helps with chemical changes, a coenzyme partners with an enzyme to produce particular reactions. Carnitine carries free fatty acids into the mitochondria, where they are used to make energy. Co Q10, on the other hand, is involved in the actual production of energy within the mitochondria.

The Co Q10 story is a saga that has been unfolding over four decades, and is still not complete. It began in 1953, when Dr. Frederick Crane, a plant physiologist at the University of Wisconsin who is known as the "grandfather" of Co Q10 research, discovered a new substance in the mitochondria of cauliflower. Since it had a yellowish color, he believed it was related to the vitamin A family, but he eventually realized it was something completely new. Four years later, he isolated the Co Q10 from beef heart mitochondria. In 1958, Professor Karl Folkers and coworkers at Merck and Co. determined the chemical structure of Co Q10, and developed a method for synthesizing it. Since Co Q10 was then very expensive to produce, and a natural substance that couldn't be patented, Merck sold its Co Q10 technology to Japanese researchers, who later developed a more cost-efficient method of producing natural Co Q10 by fermentation.

Dr. Crane soon made another surprising discovery: Co Q10 was not just in the mitochondria; it was everywhere throughout the cell. The question was: What is it doing there? To date, scientists are still trying to fill in the missing pieces.

Early on, scientists did not consider Co Q10 to be of great therapeutic importance. Drs. Crane and Folkers were the exceptions. Both men devoted much of their careers to researching the role of Co Q10 in the body. Dr. Crane, who is still involved in Co Q10 research, now says his mission in life is to spread the word about Co Q10. Until his death in 1997, Dr. Folkers also crusaded for Co Q10. In fact, he is credited with coining the word "bioenergetics" to describe the study of energy production, which he predicted would become a major field of scientific interest. He was right.

In the mid 1960s, Japanese scientists reported using a form of Co Q10 (known as Co Q7) to treat congestive heart failure. (Today, some six million Japanese take Co Q10, which is widely used as a treatment for heart disease.) Since congestive heart failure is one of the most difficult of all conditions to treat, this should have been regarded as big news. At the time, however, it received little attention. In the United States, the media were fixated on dazzling new medical breakthroughs such as open heart surgery, which showed great promise as a "quick fix" for all heart disease. Medical institutions were in a race to see who would be first to transplant a human heart. Nutrition and preventive medicine were considered inferior to the high-tech solutions found in the operating room and the sparkling new cardiac care units. Heroic medicine, not sound science, excited the American public. Only a handful of thoughtful physicians pointed out that surgery was both traumatic and very expensive or that human hearts for transplantation would be scarce, and they were ignored.

Shortly thereafter, two American scientists discovered that Co Q10 was not just instrumental in energy production, it was also an antioxidant. While this too should have been big news, it turned out to be of interest only to a small group of scientific researchers. For one thing, few understood the important role antioxidants played in the body. As I explained in Chapter 3, antioxidants protect cells against damage inflicted by unstable molecules called free radicals. If they are not stopped, free radicals can destroy all manner of cellular components, including the mitochondria, and over time will destroy the energy system. The proliferation of free radicals has been implicated in the onset of diseases such as cancer, heart disease, Alzheimer's, and arthritis, as well as other common ailments. In fact, excess free radicals are believed to be a primary cause of aging. Since free radicals are a natural by-product of energy production, Co Q10's role as an antioxidant within the energy system may not have made headlines, but it did attract extensive scientific interest.

THE HEART REVITALIZER

There has been a great deal of research performed on Co Q10, but the most dramatic results have been achieved in the field of cardiology. In 1972, Dr. Folkers and Dr. Gian Paolo Littarru demonstrated that people with heart disease are often severely deficient in Co Q10. Dr. Folkers, who was then at the Institute for Biomedical Research at the University of Texas at Austin, began clinical studies to see whether supplementing Co Q10 would help heart patients. Soon an overwhelming amount of scientific research and anecdotal evidence supported Co Q10's role in the treatment of heart disease. In some cases, its effects have been nothing less than miraculous.

If you ask cardiologist Peter Langsjoen of Tyler, Texas, whether Co Q10 is a useful treatment for heart disease, he answers with an unequivocal yes. "It's quite clear that when you add Co Q10 to other therapies, you see a very definite, easily measured improvement in the patient's quality of life—less shortness of breath and less angina. Over a period of a few months the standard measurements of heart function, whether it's by echocardiogram or nuclear scans, also show a clear improvement. Co Q10 is measurably low in heart disease, and if you supplement it, you do better. There's no question about it."

In 1981, Dr. Langsjoen and his late father, who was also a cardiologist, began giving patients Co Q10 in collaboration with Dr. Folkers. Since then, they have conducted numerous clinical trials to test the effect of Co Q10 on various forms of heart disease. As Dr. Langsjoen noted, the results have been overwhelmingly positive.

There are at least fifty studies that confirm that Co Q10 can benefit patients with heart disease. One of the best in terms of patient-group size and follow-up was conducted jointly by Drs. Langsjoen and Folkers from 1985 to 1993. It included 424 patients with various forms of heart disease who were treated with Co Q10 along with their conventional medical regimens. Patients were followed for an average of 17.8 months. All the patients were evaluated according to the New York Heart Association (NYHA) functional scale, which rates the seriousness of heart disease from I, the least serious, to IV, the most serious. Of the 424 patients, 58 percent showed improvement by moving up one NYHA classification; 28 percent moved up two classifications; and 1.2 percent moved up three classifications. Not only did the patients do better on standard cardiac tests, but many were able to stop taking some of their medication. A total of 43 percent stopped taking between one and three drugs, and only 6 percent required additional medication. Since many cardiac drugs can cause undesirable side effects such as dizziness, fatigue, nausea, rashes, and blood problems, the fact that Co Q10 eliminated the need for some drugs is particularly noteworthy. Co Q10 is not only perfectly safe but also free of side effects.

Dr. Langsjoen has become well known for his success in reversing heart failure, a condition that is not easily treated. Heart failure is the inability of the heart muscle to pump blood sufficiently throughout the body. In a very real sense, it is a disease of the energy system. People with heart failure have abnormally low levels of Co Q10, as well as high levels of mitochondrial abnormalities.

The prognosis for heart failure is often grim: although some medications may help improve heart function, they are typically stopgap measures forestalling death by only a few years or even months. The five-year survival rate is fifty percent for patients with heart failure, and those surviving for years are typically plagued with severe debility and frequent hospitalization. Moreover, this is not a disease that is going away. Although other forms of heart disease are on the decline in the United States, heart failure is actually on the rise. In 1995, according to the latest statistics available from the American Heart Association, there were more than 250,000 deaths from heart failure in the U.S.—an all-time high. Deaths from congestive heart failure have *doubled* since 1979. Dr. Folkers believed that the steady rise in death from heart failure in this country was directly due to the decline in Co Q10 which was caused by either a deficiency in the raw materials needed to produce Co Q10, or a glitch in our production mechanisms.

To Dr. Langsjoen, however, heart failure is a treatable disease, thanks to Co Q10. "I can tell you that treating heart failure with Co Q is a pleasure, because it works. I've seen patients who come here week after week, hundreds of them, who know their prognosis, they've been going downhill week by week, they've been on all the usual drugs and they're going in and out of the hospital. You put them on Co Q10 and they are often reborn, with marked improvement in measurements of heart function and a much lower need for medication."

Dr. Langsjoen's excellent results raise the question: Why isn't Co Q10 used by all cardiologists? Physicians get most of their information about new drugs from the pharmaceutical companies, who spend hundreds of millions of dollars sending representatives to every physician's office to introduce their products. Unless a new product is accompanied by free samples and a lot of hype, many physicians are suspicious of it or simply won't even hear about it.

Obviously, since Co Q10 is sold over the counter, there is little interest on the part of pharmaceutical houses to promote a substance on which they cannot make money.

Why does Co Q10 work so well? Dr. Langsjoen notes that

unlike other drugs, which merely control symptoms, Co Q10 offers physicians an opportunity for the first time in medicine to alter heart function by intervening on the cellular level. Ironically, Co Q10's very simplicity makes it difficult for cardiologists to accept. "It's too simple," says Langsjoen. "The only analogy that would be that profound and that simple would be something like this: If you were dehydrated, and I gave you water, it would be a miraculous change from the dehydrated state to the hydrated state. Giving Co Q10 to someone with heart failure produces the same kind of dramatic results."

Dr. Langsjoen feels it makes good sense not to wait until you have heart failure to take Co Q10. He believes that taking Co Q10 from midlife on (when the decline in Co Q10 begins) could have a protective effect on the mitochondria of the heart and preserve heart function.

The spectacular effect of Co Q10 on heart patients that Dr. Langsjoen has seen in his patients is elegantly illustrated by new studies that show Co Q10's effect on stressed heart cells.

MAKING OLD HEART CELLS ACT YOUNG

In a landmark series of studies, Australian surgical scientist Franklin L. Rosenfeldt, associate professor at the Baker Medical Research Institute affiliated with Monash University and Alfred Hospital, and researchers at the Center for Molecular Biology and Medicine at Monash University have clearly shown that Co Q10 can protect an old heart from injury. In one study, the researchers removed the hearts from a group of young rats and a group of old rats. The hearts were kept beating on an apparatus and allowed to circulate fluid. Under normal conditions, a rat's heart pumps around 300 beats per minute, but the researchers speeded up the heart rate to around 510 beats per minute for two hours, which was profoundly stressful. This prolonged period of stress could be lethal. After their ordeal, the young hearts were able to recover about forty-five percent of their function. The old hearts, however, recovered only seventeen percent of their function; they were performing poorly.

In the second part of the experiment, another group of animals was given Co Q10 for six weeks while a control group received a placebo. The experiment was repeated. The hearts of the animals that had not received Co Q10 performed precisely as they had in the previous experiment. The young hearts that had been pretreated with Co Q10 also recovered precisely as before. The real difference, however, was in the recovery of the hearts of the old animals: The Co Q10 fortified the old hearts. They recovered at the same rate as the young hearts! In other words, after being placed under conditions of stress, the old hearts were able to perform as well as the young hearts.

Armed with these results, Dr. Rosenfeldt decided to take the experiment one step farther. As a rule, patients over seventy do not recover as well from heart surgery as those under sixty because they suffer more acutely from reperfusion injury. During open heart surgery, oxygen and blood are cut off from the heart much as they are during a heart attack. When it is finally restored, the rush of oxygen creates an excess of free radicals, causing a great deal of damage to heart tissue. Dr. Rosenfeldt and his colleagues wanted to see if pretreatment with Co Q10 would help these patients better handle the stress of heart surgery.

In the study, surgeons collected a small portion of heart tissue that is normally removed in surgery. The tissue samples were brought back to the laboratory, cut into small strips, and placed in an organ bath that provided them with oxygen and glucose. The cells were stimulated electrically to get them to beat normally so that the force of the contraction could be monitored. For an hour, the cells were denied oxygen and glucose, which were then restored to the organ bath. Under these conditions, the younger heart muscles recovered seventy percent of their contraction strength, whereas the older muscles recovered only forty-nine percent. The researchers then repeated the same experiment, but this time putting both the old and young heart cells in contact with Co Q10 or a placebo for thirty minutes prior to cutting off the oxygen. As in the animal experiments, the addition of Co Q10 made no difference with the young muscles, but there was a striking improvement in the contraction of the old muscles. In fact, when pretreated with Co Q10, the old muscles recovered up to seventy-two percent of their contraction, slightly better than the young muscles! Already heart surgeons are administering Co Q10 to patients to improve their recovery rates.

Even though Co Q10 did not have any impact on younger heart cells in these studies, there is evidence that it may have an overall effect that combats heart disease at any age. As mentioned earlier, Co Q10 is an antioxidant. Free radical attack to lipoprotein or fat molecules can begin the cascade of events that leads to the development of atherosclerosis, or the clogging of the arteries delivering blood and oxygen to the heart. If enough Co Q10 is on hand, it can prevent the free radicals from doing their damage. Co Q10 can also block the formation of cholesterol in the body.

When combined with carnitine, the effect of Co Q10 is even more dramatic, especially in terms of maintaining healthy levels of cholesterol and other blood lipids. As noted, carnitine lowers triglycerides and raises HDL, while Co Q10 lowers overall cholesterol. At the same time, both supplements are enhancing mitochondrial function, which enables the heart to do more work with less effort. The result is a stronger, more youthful heart.

THE DISEASE-FIGHTING SUPPLEMENT

Earlier in this chapter, when I was describing the effect of carnitine on the immune system, I noted that as we age, there is a decline in immune function which makes it more difficult for us to resist disease. Both carnitine and Co Q10 have been shown to improve immune response in people whose immune systems have been suppressed due to illness. I believe that they have a similar effect on the immune systems of otherwise healthy older people.

There is no doubt that Co Q10 also increases immune function. Dr. Langsjoen, who has given Co Q10 to thousands of patients, notes that the one unsolicited comment he hears most from his patients is that they don't get sick as often. "That's a given. My patients often tell me that they haven't had a cold since they started taking Co Q10."

The mechanism by which Co Q10 enhances immune function has been the subject of much debate and speculation. Since Co Q10 is an antioxidant, it may bolster immune function by protecting immune cells from free radical attack. In addition, there has been some exciting research suggesting that Co Q10 and other antioxidants may turn on and off genes which regulate cell activity.

Although we tend to think of genes in terms of their role in heredity, in reality genes are the software that tells our cells what to do. When we are threatened by a viral or bacterial infection, our genes stimulate target organs to produce more disease-fighting immune cells. Some researchers theorize that as we age, the communication signals between cells may get crossed, and our immune cells are no longer able to react as quickly or effectively. Co Q10 may help restore immune function by restoring this communication.

A strong immune system is necessary to prevent many different diseases, from the common cold to killers like cancer. Co Q10 could play a role in both helping to prevent cancer and enhancing the effectiveness of conventional therapy. In 1991, Karl Folkers discovered that some cancer patients have lower blood levels of Co Q10 than normal, and hypothesized that low levels of Co Q10 could inhibit the body's ability to fight cancer. Several studies have shown that Co Q10 boosts the effectiveness of disease-fighting T cells, which are involved in weeding out cancer cells. In 1993, Dr. Folkers and researchers at a cancer treatment center in Denmark gave Co Q10 supplements to thirty-two typical breast cancer patients who had been designated high-risk because their tumors had spread beyond the primary site to their lymph nodes. All of the patients received surgery and other conventional treatments as well as 90 mg of Co Q10 daily. They were also given a combination of antioxidants, other vitamins, minerals, and essential fatty acids. Considering the serious condition of these women, researchers believed that there would be at least four deaths before the study was completed. In fact, none of the women died, and none appeared to get any worse over the eighteen months they were followed. Six of the women showed signs of partial remission. There have been other anecdotal reports showing some improvement in advanced cancer patients given Co Q10.

So what do these studies mean? Dr. Richard Willis, who worked with Karl Folkers at the Institute for Biomedical Research and now heads the Co Q10 research there, is cautiously optimistic: "I believe that Co Q10 helps cancer patients better tolerate conventional therapies to get the job done." He adds that Co Q10's role in cancer therapy is complementary—that is, it is meant to work in synergy with other therapies, and is not a cure on its own.

Dr. Willis adds another note of caution. Co Q10 may interfere with the effectiveness of radiation therapy but not chemotherapy. Investigations are still under way as to the role of other antioxidants in radiation therapy. **If you are undergoing radiation treatment for cancer, talk with your oncologist or physician before taking additional supplements. You may need to discontinue taking your supplements prior to and following your treatment.**

PROTECTING YOUR BRAIN

One of the newest and most exciting areas of research is Co Q10's effect on the neurodegenerative diseases, such as ALS (Lou Gehrig's disease) and Huntington's disease. There is no cure and little effective treatment for either of these diseases, which tend to strike people later in life. Much of the work on Co Q10 and the brain has been performed by researchers at Massachusetts General Hospital in Boston under the direction of Dr. Flint Beal. In these studies, older animals were given a brain poison capable of destroying the pathway that produces ATP, resulting in an animal form of neurodegenerative disease similar to Huntington's and ALS in humans. The brain cells of animals pretreated with Co Q10 did not show the telltale damage normally seen after administration of this poison. Multicenter trials are being planned to see if Co Q10 can slow down the progression of Huntington's disease in humans.

Although it has not yet been studied, researchers are hopeful that Co Q10 may have a similar effect on Alzheimer's disease.

I have no doubt that long-term studies will eventually show that the Energy Pack, Co Q10 and carnitine, has a significant impact on brain aging, as it has on aging in general.

The forty years of research encapsulated in this chapter supports what I've been seeing in patient after patient in my clinical practice. That is, the energy system is so crucial to our health that to maintain its integrity is tantamount to the best medicine that money can buy. My patients on carnitine and Co Q10 get sick less often, require fewer hospitalizations, are in better moods, sleep better, and diet more effectively.

Nature may have planned for our mitochondria to wear out and for us to conveniently disappear off the planet once we have passed our reproductive prime, but we humans have other ideas. Our bodies may be built for obsolescence, but thankfully, our brains have enabled us to find a new lease on life. The Energy Pack can revise nature's plan in midstream by reinvigorating the energy system. I have personally felt and seen the short-term benefits in myself and my patients. The Seven Signs of the Energy Crisis quickly vanish and are replaced with renewed energy, improved mental outlook, and a vitality level that is off the chart.

The long-term benefits of the Energy Pack will be even more spectacular. I believe that the *Natural Energy* program will enable us to maintain our health and vigor well beyond the age at which people are programmed to become "sick and tired" and begin the descent into illness and debility. There is simply no need to succumb to the entirely preventable diseases of aging. As I've seen time and time again in my clinical practice, once you have *Natural Energy,* you have it all!

By now you're probably excited about beginning the *Natural Energy* program. Chapter 5 will provide an overview of the program, and will give you the information you need to get started.

Chapter Five

NATURAL ENERGY: REPAIR! RECHARGE! REVIVE!

In **the last** four chapters you have seen the incredible breadth and scope of the science behind the *Natural Energy* miracle. I have taken you to the laboratories of some of the world's top scientists, who are beginning to unravel the mystery of the energy system and how it affects every aspect of our lives. I have described how the lives of my patients—and my own life—have been transformed by the *Natural Energy* program.

The program is based on my twenty years of experience treating thousands of patients, from my days managing a large ER in a bustling university center to my thriving suburban New York practice. It is supported by thousands of scientific studies performed at the world's leading research centers. It is validated by my numerous patients who have benefited from the program and who have never felt better in their lives.

Although the underlying science behind *Natural Energy* is complex, the good news is, the practice is simple—*very* simple. The program's extraordinary success is based on three easy rules: REPAIR, RECHARGE, REVIVE.

Repair your energy system with the Energy Pack, carnitine and Co Q10—two nutrients that are essential for energy production.

Recharge your body with an easy-to-follow "mitochondria-friendly" food plan that enhances the action of the mitochondria, the energy-producing powerhouses of the cell.

Revive your life by eliminating energy wasters and embracing energy boosters to keep your system running energy-efficient for years to come.

My files are filled with stories of patients who have overcome their personal energy crises with *Natural Energy.* Remember my patient John, the Sign #5 ("overeating because you're starved for energy") who reached for a candy bar every time he felt tired? John was tired a lot, and as a result, ate a lot of candy bars. He finally came to see me because he was worried about gaining weight. He had tried dieting on his own, but found that he couldn't stay awake without his frequent sugar fixes. I was worried about John because he had a strong family history of heart disease, and weight gain was just adding to his risk factors. It wasn't until he dealt with the underlying reason for his weight gain—the slowdown in his energy system—that he began to see positive results. Within a week of starting the *Natural Energy* program, John found that he was able to make it through the day without wanting a candy bar. Within four months, he had shed more than ten pounds. He felt terrific and it showed. More important, by reinvigorating his energy system, he has taken his name off my list of people most likely to wind up in the cardiac care unit!

Remember my patient Irene, the homemaker in Chapter 2 who was the victim of the "incredible shrinking day"? Run ragged by three young children, Irene was so exhausted that she had stopped doing anything to preserve her own health and well-being. True to Sign #1, her mind would scream *keep going,* but her body would frequently give out. After following the *Natural Energy* program for two months, Irene has turned her life around. Fortified by the Energy Pack, the right food, and a life free of energy zappers, Irene has found time—and energy—for the meaningful activities that she believed were lost to her forever. Six months ago, Irene could never find the time for exercise; now she's at the gym three times a week. Six months ago, she dropped out of a reading group that she loved because she couldn't keep up with the reading; now she's back. Six months ago, she couldn't imagine choosing to go out to dinner with her husband instead of going to sleep early; now the two of them make it a point to go out on Saturday nights. Irene's life is fuller and happier than it's been in years.

Remember my patient Rebecca, the forty-nine-year-old nurse who moonlighted as an aerobics instructor? Rebecca was suffering from "drop-dead fatigue." She was a clear Sign #1 if I ever saw one. Her mind was pushing her to train for a national aerobics competition, but her body was beginning to break down. Every week Rebecca would show up at my office with a different problem, ranging from backaches to headaches to knee aches to a chronic cold. I was convinced that if Rebecca continued on the same course, she would end up with a serious injury. After three months of misery, Rebecca's body wore down to the point at which she had to get some help.

A careful eater, Rebecca didn't need any help in terms of

diet. I put her on the Energy Pack, and the change in her life was palpable. But perhaps what made the biggest difference for Rebecca's life were the lessons she learned from "Revive," the third component of the *Natural Energy* program, which deals with lifestyle. Rebecca learned how to listen to her body, when to push and when to stop, and how to be kind to herself. She was finally able to train for the aerobics competition. Although she didn't take home first prize, she placed in the top ten. And she didn't fixate on energy zappers like, "Why didn't I place first?" She was proud of her accomplishment.

The experiences of patient after patient confirm the underlying philosophy of *Natural Energy:* If you tend to your energy system, your energy system will take care of you. That's the key to staying healthy and vital.

Within the first two weeks of starting the *Natural Energy* program, you will notice an appreciable improvement in energy and stamina. You will find your mood elevated and be amazed by your ability to concentrate. Over time, as the *Natural Energy* program takes hold, you will enjoy the countless other benefits that result from bolstering your energy system—from catching fewer colds to sleeping better, from fulfilling more of your goals with less effort to enjoying a life lived to the fullest. You won't be filled with regrets over missed opportunities. "I should have," "Why didn't I?" and "I can't," will be replaced with the positive "I did," "I can," and "I will."

When you have *Natural Energy,* anything and everything is possible.

Not only are the effects of *Natural Energy* palpable in terms of how you feel, but there are very real health benefits to be gained by following this program. Fortifying the energy system will enable you to postpone or even prevent the diseases that have long been associated with aging: heart disease, Alzheimer's, osteoporosis, cancer, and depression.

Slows Down Aging: Long before we see the first wrinkles or gray hairs, our cells are undergoing destructive changes that hamper all body systems. The decline begins with a slowing energy system. Without adequate energy, cells are not able to dispose of their waste products efficiently, and this allows dangerous toxins to accumulate. Without adequate energy, cells are unable to repair or replace themselves in a timely fashion. Without adequate energy, cells in key organs begin to malfunction and die. Maintaining the energy system is the only way to prevent this downward spiral.

Relieves Lethal Stress: Long-term exposure to the stress hormone cortisol can damage healthy cells and tissues throughout the body. As we age, it becomes increasingly difficult to control stress, and that itself can be deadly. The hippocampus, the memory center in the brain, is particularly vulnerable to stress-related damage. Numerous studies have linked stress to an increased risk of heart disease. *Natural Energy* helps restore the body's ability to cope with stress on two levels. First, the Energy Pack helps control excess cortisol. Second, by avoiding energy zappers, you are mitigating the stress in the first place.

Revitalizes Your Heart: Heart disease is very much a disease of the energy system. Rich in mitochondria, the heart is one of the most energy-dependent organs in the body. When the mitochondria slow down, the heart cannot pump as efficiently. It has to work harder and harder to do less and less. The slowdown in mitochondria also prevents fats from being metabolized properly, which leads to atherosclerosis (clogged arteries). The net effect of the energy crisis is to weaken heart function and leave us vulnerable to heart attacks. The Energy Pack can turn back the clock by rejuvenating heart mitochondria.

Protects Your Brain: The brain is one of the most metabolically active organs in the body. Long-term exposure to stress hormones and toxins produced through normal energy production can destroy brain cells. Your brain's only hope is a strong energy system which can eliminate toxins efficiently and repair damaged cells before they die.

Boosts Your Immune System: As your body ages, your immune system ages too. It becomes less efficient at fighting viruses and bacteria and keeping cancerous cells contained. The problem is, your body is expending so much energy keeping all systems running and doing the necessary repairs that there is little left in reserve to fight off disease. As a result, we succumb to the diseases of aging. *Natural Energy* will boost immune function by restoring youthful health and resilience.

Prevents Weight Creep: Changes in lifestyle and metabolism result in "weight creep," the constant weight gain that starts in midlife. Beginning around age forty, the average person may gain up to ten pounds a decade. No wonder recent studies show that up to half the adult U.S. population is officially overweight! The *Natural Energy* program can combat weight creep by ending the destructive "eat too much, do too little" lifestyle that leads to excess weight.

Although the *Natural Energy* program is general enough to be followed by almost everyone, I have included specific advice for those of you who identify particularly with one or more of the Seven Signs of the Energy Crisis. By incorporating the additional advice accompanied by your icon into the general program, you will be able to tailor your program to meet your particular needs.

Before you begin the *Natural Energy* program, you should understand the whole picture.

Repair

The "Repair" component of the *Natural Energy* program stops the energy crisis on the cellular level. Taking the Energy Pack jump-starts your energy system by giving your mitochondria what they're craving. The Energy Pack is easy to use—all you have to do is swallow two little pills with your meals. Yet its effects on the energy system are far-reaching. The centerpiece of the *Natural Energy* program, the Energy Pack alone can halt and reverse the energy crisis. I'm confident that in just a little while, you'll be amazed at the change that's come over you.

As you know by now, our mitochondria/energy factories break down as we age. To compound the problem, levels of the key nutrients needed to make energy—carnitine and Co Q10—begin to drop as well. The net result is devastating. We must work harder to make less energy, even while our bodies are facing more and more challenges. Not only must we have energy for day-to-day operations, but as our bodies age, old cells need to be repaired or replaced; we need an extra boost to maintain the cellular machinery. Moreover, we need deep reserves of power to heal when major things break down, as they inevitably do.

The Energy Pack helps rejuvenate tired mitochondria and restore youthful energy production. By increasing levels of these key nutrients, we give our mitochondria what they need to act like kids again.

The Energy Pack is what *Natural Energy* is all about—giving your body what it wants and needs to function at optimal levels. Both carnitine and Co Q10 are naturally produced by the body, but levels decline with age. We need to take supplements so that our cells can make energy again.

There is no doubt that carnitine and Co Q10 can be of great benefit to people who are already in the throes of the energy crisis, and are suffering from its consequences. Exciting new evidence shows that early intervention—beginning the Energy Pack before the energy system takes a dive—may help keep it from breaking down in the first place. A recent study suggests that chronically low levels of carnitine could, over time, cause all manner of destructive changes in heart cells that could lead to heart disease. People with heart disease, cancer, and other serious illnesses often have low levels of Co Q10. In the case of heart failure, Co Q10 supplementation has been nothing short of miraculous.

To my way of thinking, why wait for the energy system to break down before taking action? I've seen in my clinical practice that maintaining the cellular machinery with the Energy Pack will help prevent it from falling apart in the first place.

In Chapter 6—"Repair!"—you will have a choice of following a basic energy-enhancing program or one specifically tailored to one of the Seven Signs of the Energy Crisis. Whatever program you choose, you will be taking the Energy Pack, albeit with varying doses of carnitine and Co Q10 to accommodate individual needs. Depending on your sign, you will need to take your supplements up to four times daily, with meals or a snack.

If you are following the program designed for one of the Seven Signs of the Energy Crisis, I recommend taking additional supplements which are targeted to specific problems and will enhance the effect of the Energy Pack. You will learn who should take them and how they should be taken. Similar to the Energy Pack, these additional supplements are completely safe, and are sold over the counter at the same stores where you can buy the Energy Pack.

As important as the Energy Pack and the energy-boosting supplements may be in terms of pumping the energy system, they are not the whole story. They work best when incorporated into the entire *Natural Energy* program. When a patient is sick and requires an antibiotic, I do not simply write a prescription and send him on his way. Illness is a sign that the body needs special pampering. I tell my patients that it is equally important to give their bodies a chance to heal by getting enough rest, drinking extra fluids, eating well, and steering clear of stress. This formula works like a charm. The same is true for treating the energy crisis: it is not enough to merely take your Energy Pack and expect to bounce back overnight. You must take care of the whole body—from your cells on up. Otherwise, another energy crisis is sure to result.

Recharge

What you eat, when you eat, and why you eat is critical to maintaining the energy system. Food provides the raw material from which our cells produce fuel. The right food can help the system run reliably and smoothly, but the wrong food can clog it and hasten its demise.

What's good for your energy system, it turns out, is also good for every other system in your body. *Natural Energy*'s mitochondria-friendly food plan gives your cells the materials they need to maximize energy production. In addition to revving up the energy system, *Natural Energy*'s food plan will provide powerful protection against disease long associated with poor nutrition, such as cancer and heart disease. Perhaps as many as half of all cases of cancer and heart disease are due to poor diet!

But now that I have your attention, let me put your fears to rest. I'm not suggesting that you survive on seeds, bran, and weeds. Nor I am going to have you count calories and fat grams, or measure out your food morsel by morsel. Healthy, high-energy eating doesn't have to be dull, boring, or weird. You will be delighted with the wide variety of food you can eat and still stay within the mitochondrial-friendly zone. As you'll see, it's as much when you eat as what you eat.

The point of "Recharge" is to make you aware of what you eat without making you neurotic. Although I offer general guidelines for high-energy eating, I do not insist that you adhere to a strict diet. Once you become aware of the role food plays in how you feel, you'll embrace foods that enhance the effectiveness of the *Natural Energy* program, and banish energy-draining eating.

A strong link exists between fatigue and hunger, both in how it affects the biochemistry of the body and in how our brains interpret those signals. It is easy to mistake the light-headedness, tiredness, and irritability associated with fatigue for signs of hunger. When you eat because you're tired, it can actually make you more tired, which of course will only make you think that you want more food. Eating energy-boosting foods at the right times, as I discuss in chapter 7, will help get you off this vicious circle of hunger and exhaustion. Although "Recharge" is important for everyone who wants to achieve *Natural Energy,* obviously it is of critical importance for those of you with Sign #5, who overeat because you're overtired. Solving your personal energy crisis is the *only* way to get yourself off the diet roller coaster.

In "Recharge!" I offer general advice about nutrition that applies to everyone, but remember to look for your sign. There are plenty of tips for those of you with one or more of the Seven Signs of the Energy Crisis. I will also show you how easy it is to put together a tasty high-energy meal or snack in virtually no time.

Revive!

While the Energy Pack and the mitochondria-friendly food plan provide your cells with the right raw materials needed to make high-quality energy, "Revive" focuses on reinvigorating your energy system from the outside in. As important as improving your internal environment may be, the wrong external environment can rapidly deplete you of your newfound energy. There is no point in having more energy if you are going to simply fritter it away on energy zappers that can rob you of your quality of life.

My work with thousands of patients has taught me that good medicine extends far beyond the treatment of physical illness. It should also include an examination of emotional factors that may have contributed to—or even triggered—the illness in the first place. It is not enough to reinvigorate an ailing body; it is equally important to revitalize an ailing spirit. I have learned that if you don't get the body and mind in sync, there is a good chance that one will bring the other down.

Natural Energy is not just about a strong body and a sharp mind. It is about a renewal of the spirit and a recapturing of the joy that makes life worth living. "Revive!" shows you how to make your spirits—and your energy levels—soar.

First things first. It's time to get your energy chemistry back in shape. So hold on. You're about to feel like a kid again!

Chapter Six

REPAIR!

Less than an hour after I prescribed the Energy Pack to Andrea, she called my office from outside a nearby health food store. She was flustered and frustrated, and sounded like she was about to break into tears. "Erika," she said, "this place is so confusing. I can't find anything!"

I had forgotten how baffling health food stores can be for people who aren't used to them. I can usually find what I need in a few minutes and be on my way, but I've been shopping in them for twenty years. It turned out that this was the first time in her forty-two years that Andrea had ever tried to buy anything at a health food store. Now Andrea is about as smart and sharp as they come, but she was accustomed to handing a prescription to her pharmacist and getting what the doctor ordered in return. The task of navigating her way through row upon row of unfamiliar labels, identifying the right product in the correct dosage, was more than she could manage.

Fortunately, I had a break between patients. I met Andrea at the store and actually walked her up and down the aisles explaining some of the product labels and showing her how to find what she needed.

In the process of showing Andrea around the store, I learned as much from Andrea as she learned from me. It is a lesson that every physician who prescribes over-the-counter supplements should take to heart. Before we turn our patients loose in the supplement aisles, we need to make sure we have armed them with the knowledge they need to purchase and use these supplements correctly. That is the purpose of this chapter. I will guide you through the process of buying the Energy Pack and using it. I will tell you where you can buy these supplements, how to select a high-quality product, and how you should take them.

The first thing you should know is that supplements are available virtually everywhere, from health food stores to drugstores to supermarkets to discount department stores. They are even sold on the Internet. As I will discuss in greater detail later, buying a supplement is no different from buying any other product. Your best bet is to stick to name brands from reputable manufacturers.

Most of you, however, will probably be buying your supplements at health food stores. Each store tends to be set up a bit differently, and has its own way of organizing its product lines. For example, some stores may sell carnitine in a section with other amino acids; others may sell it in the vitamin section. Some stores may sell Co Q10 in a special section devoted to enzymes; others may sell it along with vitamins. If you are unfamiliar with the layout of a store, or can't find what you're looking for, simply ask the salesperson or manager to direct you to the carnitine and Co Q10.

Never on an Empty Stomach

Take your Energy Pack with food and a full glass of water. Some people may find that taking supplements on an empty stomach causes stomach distress. Food not only helps to buffer the supplements so that they're easier on the stomach, but aids in the absorption of these nutrients by the body.

I can't overstate the importance of water. It helps to speed up absorption of the Energy Pack, but it also performs other vital roles. Most of us do not drink enough water to adequately maintain the cells and tissues of our body. We live in a state of near dehydration. Remember, the human body is more than seventy percent water, and we need to replenish our water stores frequently. In addition, many of the beverages that we drink contain natural diuretics such as caffeine which further diminish our water supply. Drinking a full glass of water every time you take your Energy Pack is a wonderful way to ensure that you are adequately hydrating your body.

Which Program Is for You?

Review the Seven Signs of the Energy Crisis. If you identify strongly with one of them—if the first thought that pops into your mind is, "That's me!"—follow the program accompanied by your sign. Some of you may say yes to several signs. In this case, you may have to do a bit of soul-searching to focus on your main problem. If you can't narrow it down to one specific sign, follow the general program.

For an Added Boost

At times, I recommend additional supplements for those of you who need a special boost. If you have one or more of the Seven Signs—for example, Sign #1 and Sign #5—you may add the additional supplements listed for those signs to your basic program. If you pick two signs with contradictory dosage information for carnitine or Co Q10, always rely on the higher doses. The extra boost you need for one sign won't get in the way of the other.

All of the supplements that I recommend are meant to work together with the Energy Pack. They are safe and sold over the counter at health food stores and pharmacies.

Don't Leave Home Without Them

The basic Energy Pack program requires taking supplements twice daily, once in the morning and once at night. If you follow a specialized Energy pack program, you may have to take your supplements up to three times daily. For most of you, that will mean carrying your midday supplement supply with you. You can carry your day's supply of supplements in a handy pill case, or keep an extra supply at work or in your car. On occasion, you may skip a dose by accident. Don't worry about it; simply pick up where you left off the next time. For the program to work, however, you need to take your supplements daily and consistently.

Natural Energy for Everybody
The Basic Plan

If you are comfortable taking supplements, and strongly identify with one or more of the Seven Signs, skip the basic plan. Follow the program that best suits your needs under "Natural Energy for the Seven Signs" below.

If you find that you identify with all of the Seven Signs, consider the basic plan first. It may give you the energy boost you need, or at least stabilize your energy system enough for one of the signs to become of paramount concern.

BREAKFAST

Take with breakfast and a full glass of water.

Carnitine	500 mg
Co Q10	30 mg

LUNCH

Take with lunch and a full glass of water.

Carnitine	500 mg
Co Q10	30 mg

Natural Energy for the Seven Signs

If you identify with any of the Seven Signs, follow the program listed below that best suits you:

Hearing your mind say "now" when your body says "later."

The goal for Sign #1 is to give your body a general energy boost so that your body and mind can be in sync. When your mind says "Go," soon your body will say, "No problem!" Along with the Energy Pack, I recommend that Sign #1 also take magnesium, a mineral critical for energy production but often deficient in our diets.

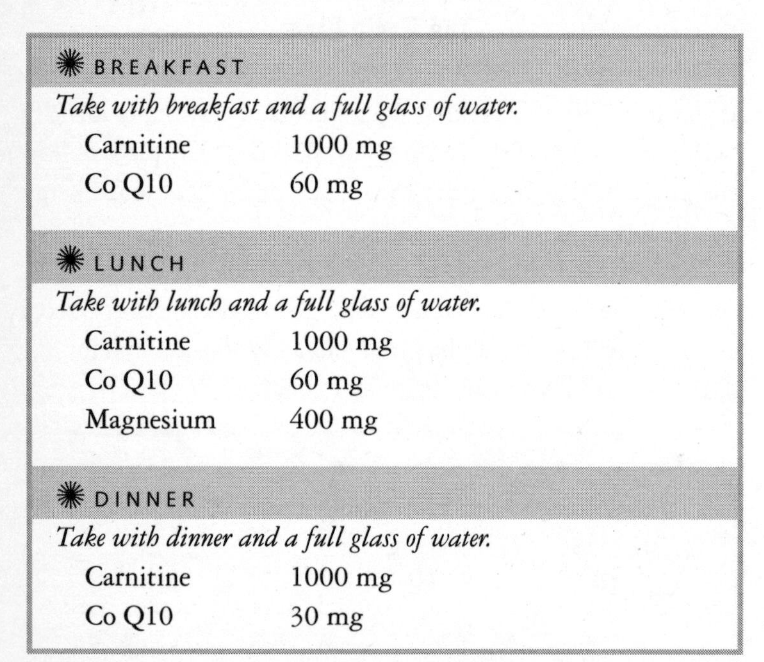

BREAKFAST	
Take with breakfast and a full glass of water.	
Carnitine	1000 mg
Co Q10	60 mg

LUNCH	
Take with lunch and a full glass of water.	
Carnitine	1000 mg
Co Q10	60 mg
Magnesium	400 mg

DINNER	
Take with dinner and a full glass of water.	
Carnitine	1000 mg
Co Q10	30 mg

Waking up more tired than when you went to sleep.

People who wake up DOF (dead on their feet) need an extra energy boost in the morning, which is why I recommend additional magnesium and calcium with breakfast. Be sure to follow my tips on eating in Chapter 7, "Recharge!" The right combination of food and supplements can make your mornings a lot brighter.

BREAKFAST	
Take with breakfast and a full glass of water.	
Carnitine	1000 mg
Co Q10	60 mg
Calcium	500 mg
Magnesium	400 mg

MIDMORNING

Take with a high-energy snack (see Chapter 7) and a full glass of water. After two weeks, you can discontinue this extra dose.

Carnitine	500 mg
Co Q10	30 mg

LUNCH (OR DINNER)

Take with lunch and a full glass of water. Do not drink any caffeinated beverages after eleven a.m. They could disrupt your sleep, leaving you even more tired the next morning.

Carnitine	1000 mg
Co Q10	60 mg

Yearning for a midafternoon nap.

The afternoon droop is one of the most common signs, but it's one of the easiest to fix. This plan will make your batteries stay charged all day long. For the first two weeks, you will need to take an additional after-lunch dose with a high-energy snack.

BREAKFAST

Take with breakfast and a full glass of water.

Carnitine	500 mg
Co Q10	30 mg

LUNCH

Take with lunch and a full glass of water. Be sure to avoid energy-zapping foods for lunch (see Chapter 7).

Carnitine	1000 mg
Co Q10	30 mg

MIDAFTERNOON

Take with a full glass of water and a high-energy snack. After two weeks, you can discontinue this dose, but continue eating the snack.

Carnitine	500 mg
Co Q10	30 mg

Walking around in a brain fog.

When your ability to concentrate and think clearly is impaired, it is a sign that the energy crisis hits you hardest right between the ears. In addition to the basic Energy Pack, I recommend that you take two additional supplements: magnesium and L-glutamine. L-glutamine is an amino acid that helps to stabilize blood sugar levels. This prevents the sharp dips in blood sugar that can contribute to brain fatigue.

BREAKFAST

Take with breakfast and a full glass of water. A cup of coffee may help bring the day into focus, but avoid fruit and sugar-laden foods.

Carnitine	1000 mg
Co Q10	60 mg
Calcium	500 mg
Magnesium	400 mg
L-glutamine	800 mg

LUNCH

Take with lunch and a full glass of water.

Carnitine	1000 mg
Co Q10	60 mg
L-glutamine	400 mg

DINNER

Take with dinner and a full glass of water. Avoid alcohol; it will further impair your ability to think sharply.

Carnitine	500 mg
Co Q10	30 mg
L-glutamine	400 mg

Overeating because you're starved for energy.

Exhaustion and hunger have one thing in common: they are both triggered by a rapid drop in blood sugar. As you know, that sudden hollow feeling can often lead to overeating. In addition to the Energy Pack, I recommend taking magnesium, L-glutamine, and calcium. Calcium works with L-glutamine to inhibit the biochemical chain reaction that leads to the sugar highs and lows.

BREAKFAST

Take with breakfast and a full glass of water.

Carnitine	500 mg
Co Q10	30 mg
L-glutamine	800 mg
Magnesium	500 mg

LUNCH

Take with lunch and a full glass of water.

Carnitine	1000 mg
Co Q10	60 mg
L-glutamine	800 mg

LATE AFTERNOON (AROUND FIVE PM)

Take these additional energy boosters about two hours before dinner. They will help to keep your blood sugar stable. You can discontinue this dose after two weeks.

L-glutamine	800 mg
Magnesium	400 mg
Calcium	500 mg

Feeling stressed out and blue.

Low energy depletes the body and the mind. When we're suffering from "hit the wall" exhaustion, our ability to cope flies out the window. We are left depleted from head to toe. In order to lift your mood, you need to boost your mental and physical energy. In addition to the Energy Pack, I recommend taking magnesium and L-glutamine, for all the reasons I've explained above, plus two new supplements, vitamin C and omega-3 fatty acids. Vitamin C enhances immune function, which is often depleted during times of stress. Omega-3 fatty acids are essential for normal brain function, and low levels can promote depression.

BREAKFAST

Take with breakfast and a full glass of water.

Carnitine	1000 mg
Co Q10	60 mg
Magnesium	400 mg
Omega-3	1000 mg

LUNCH

Take with lunch and a full glass of water. Try to eat some foods rich in omega-3, such as salmon, albacore tuna, almonds, avocados, or grains, for lunch.

Carnitine	1000 mg
Co Q10	60 mg
L-glutamine	800 mg

DINNER

Take with dinner and a full glass of water. Guard against using alcohol as a means to lift your spirits.

Carnitine	1000 mg
Co Q10	60 mg
L-glutamine	800 mg

Fearing that your sex life is stuck in low gear.

Any or all of the Seven Signs of the Energy Crisis can culminate in Sign #7. The same energy zappers that cause Signs #1 through 6 will dampen your desire for sex. To the basic plan, I add magnesium for added energy, and omega-3 fatty acids to put you in a better mood, which will then put you in *the* mood.

BREAKFAST

Take with breakfast and a full glass of water.

Carnitine	1000 mg
Co Q10	60 mg

DINNER

Take with a light dinner and a full glass of water.

Carnitine	1000 mg
Co Q10	60 mg
Omega 3	1000 mg
Magnesium	400 mg

The Energy Pack

The basic Energy Pack contains carnitine and Co Q10. Both of these supplements are sold in different forms, as I describe below.

There are no contraindications to taking either Co Q10 or carnitine—I have found both nutrients to be completely safe even when taken with other prescription medications. However, I prefer to err on the side of caution. **If you are taking any drugs for an existing medical condition, check with a physician before using any supplements.**

CARNITINE

Carnitine is sold primarily in two forms—acetyl-l-carnitine and L-carnitine. What is the difference between the two? In the body, L-carnitine is actually converted into acetyl-l-carnitine. Theoretically, acetyl-l-carnitine is better absorbed into the bloodstream because it is closest to what the body uses. However, for a basic energy boost, I have used both forms of carnitine interchangeably and do not feel there is a significant difference. Both are excellent products.

Carnitine is sold in capsules, pills, and liquid form. There are several different brands of carnitine. As I mentioned earlier, whenever you purchase supplements, you should stick to reputable, well-known brands.

A month's supply of carnitine will cost you around sixty dollars, or slightly more. Some of the larger chains and mail order houses may offer discounts or on occasion have two-for-one sales. As with anything else, if you shop around you can save yourself some money, but don't save a few bucks at the expense of your health.

Most of you will have to cover this expense out of pocket, but in my opinion, it is well worth it. However, carnitine is also available by prescription under the brand name Carnitor (330 mg tablets) and has been approved by the FDA for treatment of specific carnitine deficiency diseases. If after reviewing the scientific literature, your doctor feels that carnitine supplementation is important for your health, he or she can prescribe Carnitor and that may make it reimbursable, depending on your health insurance.

Carnitine is usually sold in 250 mg, 400 mg, or 500 mg capsules.

CO Q10

Co Q10 is available in many different forms, but I recommend the micronized, hydrosoluble soft gel capsules, which are most easily absorbed by your body.

Co Q10 is sold in 15 mg, 30 mg, 60 mg, 75 mg or 100 mg soft gels and capsules.

The Other Energy Boosters

Carnitine and Co Q10 are the primary supplements I recommend for restoring the body's energy system. In some cases, however, depending on your sign, I have added an additional supplement or two to work with the Energy Pack. Below I describe the four energy boosters that can tailor the Energy Pack to your needs.

MAGNESIUM

Magnesium is a mineral essential to the energy-producing cycle within the body. Like Co Q10, it is part of the reaction that produces ATP. Without it, you can't make ATP. Since our bodies do not produce magnesium, it must be obtained from foods we eat such as wheat bran, almonds, fruits, fish, and legumes. Modern food-processing techniques often strip magnesium from food, so unless you eat a diet consisting of fresh unprocessed foods, you may not be getting enough magnesium.

Since most of us eat a diet that is far from ideal, mild magnesium deficiency is fairly common, especially in the United States. It is a particular problem for diabetics, people who regularly drink alcohol, those who use diuretics (including caffeine), and people with heart disease. In fact, low levels of magnesium may be linked to two conditions normally associated with exhaustion: PMS and chronic fatigue syndrome. In both cases, magnesium supplements have proven helpful. Magnesium is particularly important for normal heart function, regulation of blood sugar, and creation of bone.

People with kidney problems should avoid magnesium supplements.

L-GLUTAMINE

L-glutamine is an amino acid that helps to stabilize blood sugar levels. Chronic overeaters or those with Sign #5 (overeating because you're overtired) often suffer from wildly fluctuating blood sugar levels. These swings usually stem from a sudden release of insulin, which causes blood sugar to plummet and leaves you feeling like you're running on empty. Sugary and starchy foods are notorious for causing major insulin releases, and that is why some people can become ravenous an hour or two after eating a big meal, especially if it is high in such foods. Needless to say, sugar swings can also leave you feeling depleted and exhausted. By moderating the sugar highs and lows, L-glutamine will help curb overeating by preventing those energy lows that send you running for a candy bar.

CALCIUM

When they think of calcium, most people think of strong teeth and bones. In fact, calcium is a mineral that plays many different roles in the body. I recommend it for overeaters because it can help relieve cravings. When loaded with food, the stomach sends a message to the brain to release insulin to break down the sugar load. When insulin levels go up, blood sugar levels go down. This cycle leaves you hungry, tired, and craving more sugar. Calcium can stop the chain of events early on in the cycle, before the real damage is done. Recently, calcium has been shown to be an effective treatment for the cravings and mood swings associated with PMS.

OMEGA-3 FATTY ACIDS

There's a good reason why fish has always been known as brain food. The fact is, it contains high levels of the same kind of fat—omega-3 fatty acids—that makes up much of your brain cells. Omega-3 fatty acids are found in fatty fish (sardines, mackerel, salmon), whole grains, and seeds. Over the past fifty years, the consumption of foods containing omega-3 fatty acids has measurably declined in the United States. At the same time, the frequency of depression has skyrocketed. This is not just a coincidence, according to researchers at the National Institute of Mental Health. In fact, these scientists believe that if the brain is deprived of omega-3 fatty acids, this will trigger depression in susceptible people. I recommend omega-3 fatty acid supplements for anyone who is feeling down. Another reason to cheer up: Omega-3 fatty acids may also protect against heart disease and cancer.

Answers to Your Questions About Supplements

WHERE CAN I BUY SUPPLEMENTS?

As noted earlier, the Energy Pack and other supplements are sold at health food stores, pharmacies, and supermarkets, and even on the Internet. You can purchase your supplements wherever it is most convenient and economical for you.

Nutritional supplements are not regulated by the government, so to be sure you are getting the best-quality products, stick to brands from reputable, well-known manufacturers that take special steps to ensure safety and effectiveness. None of the supplements that I recommend are dangerous in any way, but there is a risk that some unscrupulous manufacturers may water down a product so that it does not contain the quantity of supplement that it should. Most of the well-known manufacturers, however, have good quality control. Look for products that come in tamperproof packages with both an inside and an outside seal. As a rule, seek out products that state on the label that they are laboratory-tested and guaranteed. That means the contents have been assayed by an independent laboratory. To be sure the product is fresh, look for an expiration date on the package; an old supplement does lose its effectiveness. Also look for a quality-control number on the package. On the off chance that something is wrong, the manufacturer can quickly pull the product off the shelf. It is another sign of well-made supplements.

Most supplements do not come in childproof packages because the capsules and tablets are difficult enough to swallow to generally deter children. In any case, supplements tend not to be poisonous even at high doses. However, take nothing for granted. If there are children in your home, keep your supplements out of their reach.

SHOULD I TAKE A MULTIVITAMIN?

I'm not a big fan of multivitamins. They go against my underlying philosophy that everybody is different. Instead of giving the same pill to everyone, I believe it is wiser to selectively take the supplements that you need at precisely the right dose for you. Typically, standard multivitamins contain very low doses of a wide range of vitamins and minerals. The doses are too low to be effective, yet they create a false sense of security that you have covered all your bases. This complacency can lead to inattention to your diet.

First and foremost, it is important to get as many vitamins, minerals, and nutrients through food as possible. There is simply no substitute for good nutrition. However, when it is impossible to obtain enough of some nutrients through food, as is the case with carnitine and Co Q10, then the strategic use of supplements can be a true lifesaver.

CAN I OVERDOSE ON SUPPLEMENTS?

I know many people believe that if a little is good, then a lot must be better. This is not necessarily the case. Your body cannot absorb a huge amount of anything at one time. There is no benefit to taking too much of any supplement in one dose; chances are your body will simply excrete the excess. In the case of some supplements—vitamins A and D, for example—megadoses can actually harm you.

As far as safety is concerned, the supplements that I recommend are nontoxic even at relatively high doses. I can't make the same claim for many prescription and over-the-counter drugs.

However, this shouldn't give you carte blanche to take as much as you want. Bombarding your body with an excess of anything is foolish. The doses that I recommend have been thoroughly researched and designed for maximum effectiveness. There is no reason to take more.

MY DOCTOR SAYS THAT SUPPLEMENTS DON'T WORK. WHY ARE DOCTORS SO RESISTANT TO PATIENTS TAKING SUPPLEMENTS?

Many physicians are still resistant to the concept of preventive medicine in general and supplements in particular. Physicians are trained primarily to treat disease: At no time in medical school are we ever taught that the maintenance of *health* should be the primary goal of the practice of medicine. We are taught instead to fight the diseases that have already appeared. And we are taught to prescribe pharmaceutical drugs. In fact, most physicians rely on representatives from drug companies to keep them updated on the latest treatments. Clearly, the system is stacked in favor of waiting until a patient gets sick and then treating the illness with drugs.

Supplements, on the other hand, work best when they are used to prevent disease. They can fortify the body so that it can keep illness at bay, and they can even help the body shake a minor ailment such as a cold or benign virus. Since drug companies do not push supplements (if they manufacture them at all), doctors still remain surprisingly ignorant of these powerful medicines.

WHEN DO YOU TREAT A MEDICAL PROBLEM WITH A SUPPLEMENT AND NOT A DRUG?

I'm often asked this question, and the answer leads back to another question: When do you call your doctor, and when do you try to treat a problem on your own with natural remedies? **If you are suffering from alarming symptoms—if you are burning up with fever, are vomiting, are dizzy, have chest pains, are incoherent, or are showing other signs of serious illness—you must contact your physician immediately. Do not attempt to self-medicate with supplements or any other over-the-counter or prescription medication without your doctor's guidance.** If, however, you are feeling run-down but your symptoms are not dire, instead of assuming you are sick, know that your body is sending you a message. Listen to what it is saying. It is probably telling you to slow down, get more rest, and drink more fluids. Help your body to heal itself by taking immune-boosting supplements such as vitamin C or the herb echinacea. If within three days you do not feel better, or you develop new symptoms, make sure you call your doctor.

WHY AREN'T SUPPLEMENTS APPROVED BY THE FDA?

The clinical studies required to get a new drug approved by the FDA can cost up to a billion dollars. Since supplements cannot be patented, there is little incentive for a pharmaceutical company to invest money that it will not recoup through a patent-protected product. An exception is carnitine, which is approved by the FDA and also sold over the counter.

The fact that a supplement has not undergone FDA approval does not mean it is untried or dangerous. The supplements that I recommend are part of the very structure of our bodies. On the other hand, many drugs that have undergone the FDA approval process not only have serious side effects, but in some cases have proven to be so dangerous that the FDA and/or the manufacturer have pulled them off the market.

IF I'M UNDER TREATMENT FOR CANCER, CAN I TAKE THE ENERGY PACK?

When your body is fighting disease, it needs an energy boost so that it can concentrate on getting well while at the same time continuing to function normally. Both carnitine and Co Q10 have been shown to either enhance the effectiveness of conventional cancer therapies or reduce some of the side effects associated with strong chemotherapeutic drugs. Other supplements may help as well. If you're interested in taking the Energy Pack or other supplements, check with your oncologist or a nutritionally oriented physician.

As noted in Chapter 4, if you are undergoing radiation treatment for cancer, keep in mind that Co Q10 may interfere with the effectiveness of the treatment. You will need to discontinue taking Co Q10 during and following the treatment period. Do not resume taking Co Q10 until your physician says it's okay.

So that's it: the keystone of the *Natural Energy* program. In just a few days, you will be amazed at the change that's come over you. But this isn't all. In the next chapter, I will show you how eating the right food can make your energy levels soar, and how eating the wrong food can crash your energy system.

Chapter Seven

RECHARGE!

Natural Energy is about greater vitality, increased mental focus, and a brighter outlook on life. In the short run, it is about looking and feeling terrific—not just on good days, but every day. In the long run, it is about preserving the body's master system—the energy system on which every other system in the body depends.

I've shown you how the mitochondria need more help in producing our cellular fuel—ATP—as we age. I've also described how the running down of the energy system can fast-forward the aging process and lead to cellular breakdown and disease. The Energy Pack helps to reinvigorate mitochondria by providing the natural boosters needed to make ATP. But the Energy Pack cannot do the job alone. It is meant to work in combination with the right energy-enhancing food.

This isn't a prescribed food plan with daily menus and odd combinations of strange foods. Rather, these are general guidelines you can follow, making sure that over the course of a few days or weeks, you're giving your body the energizing foods that help it run best.

Good Food Makes Great Energy

In order to understand why the right food is so important, you need to know a bit about how your body uses food. People often think of food as fuel, but that's doing a disservice to your body; it is not a simple machine like an automobile that gets its fuel pumped in and has little to do with the process. Your body has to make its own fuel, which it does from food. Before food can be used by the body, it must be broken down into smaller components that become the raw ingredients used to make energy. This process is called metabolism.

Your body treats food in much the same way as a refinery turns crude oil into high-octane gasoline. Only after the food has been fully refined can the cells "burn" the fuel to make energy. You've heard the expression "junk in, junk out." This is particularly true when it comes to food. The wrong food can clog the energy system and leave you feeling drained, exhausted, and surprisingly, hungry. The right food can enhance energy production and leave you feeling satisfied, healthy, and recharged.

Too often our diets emphasize convenience over quality, but I don't believe we should sacrifice one for the other. The mitochondria-friendly food plan not only will give you the tools to maximize energy production, but will let you do so in an easy fashion. In addition to boosting energy, following these simple guidelines will dramatically reduce your risk of developing many common diseases, from heart disease to cancer to diabetes to obesity.

Keeping in mind that good food leads to great energy, you'll find it easier to choose those which contribute to your overall sharpness and stamina. You'll also be more motivated to avoid the energy-sapping foods—high-sugar, processed foods like candy and soda—that can clog the body's energy system and cause weight gain and a sluggish metabolism.

From Food to Energy

Every mouthful of food we eat is broken down into smaller components in the digestive tract, and then broken down yet further into a form that can be utilized by the cells to produce energy. Whether it's a bowl of cereal, a candy bar, or a hamburger, the body reduces whatever foods we eat into three basic substances—glucose (simple sugar), amino acids (the building blocks of protein), and fatty acids (the building blocks of fats). These broken-down foodstuffs are then transported across the cell membrane into the mitochondria to make energy and all of the chemicals necessary for life.

The process of metabolism produces waste products or toxins that are excreted from the cells. This is inevitable; it's a byproduct of being alive. However, many of the foods we eat in our modern diet, particularly those heavily processed foods that are high in artificial, unnatural, and unhealthy additives, produce high levels of toxins. The more toxins produced, the more energy the cell expends to remove them. Less power is left for running the body. And the built-up toxins themselves can injure mitochondria, which can further slow down energy production, which leads to ever greater waste accumulation, and so on. The goal of the mitochondria-friendly food plan is to give the body those foods that are most efficiently, naturally, and cleanly converted into glucose, amino acids, and fatty acids.

Think Like a Cell

Our cells are obsessed with only one goal: surviving. They are simple creations, part of the large assembly line of the body, with no intelligence and no ability to distinguish between the foods we eat. In fact, no matter what we eat, our cells, in their profound simplicity, always look to the food for the same three things: glucose, amino acids, and fatty acids. Whether you eat the freshest salad greens or the heaviest piece of meat, your enzymes jump to attention the minute they receive food, and work as hard as they have to in order to break it down into these three basic substances.

Our cells aren't too bright. They not only cannot distinguish between the types of food we ingest, they behave as though each morsel of food they receive will be their last. They can't imagine there might be another meal after the one they're working so hard to digest. So they don't ease up. They are compelled to extract every last drop of nutrition from every foodstuff that comes their way. They don't want to let any potential source of energy leave the body until they've gotten every single nutrient.

The only way to keep your cells from overworking and wasting energy in their quest to make energy is to work with them. This means adapting to their ways, going on their schedule, and working with their methods. You wouldn't put sludge in your gas tank, because the engine would have to push well beyond its limits to extract whatever usable fuel might be there. Why take in foods that make your body sluggish? If the cells spend all their energy trying to break down a heavy meal into basic foodstuffs, there's not a lot of energy left to think or move.

So in order to help your cells do what they do best, you have to give them the foods they can use most efficiently. When your cells are trying to make useful foodstuffs from foods that don't have any real use, they become exhausted, and you're bound to feel tired all the time. There will be less energy available for life's pleasures when more energy is used up simply trying to survive.

Well-Fed Cells Have Energy to Spare

Though our cells work relentlessly to extract energy from food in order to keep themselves alive, that's not an end in itself. Cells stay alive so they can produce excess energy to keep our organs alive. This is their big work—keeping the heart pumping, maintaining balance in the nervous system, oxygenating the entire body via the lungs, digesting and absorbing food, and so on.

Eating What Comes Naturally

The modern age has given us all kinds of conveniences, and the world of food is no exception. Advances in food processing allow a chicken that has been raised at one end of the country to be shipped to a restaurant thousands of miles away. Grain products can be boxed to sit on grocery shelves for months. Canned foods can last for years. The problem is that to achieve these modern miracles, the foods are laden with preservatives.

Our cells have no innate ability to deal with the processed, refined, and artificial foods that make up so much of today's standard diet. Though our taste buds have been corrupted to the point at which we can find these unnatural foods palatable, our cells are not satisfied. They still do their best job taking the nutrients from natural food substances, but they can digest only foods for which our bodies produce enzymes. We don't have enzymes for the chemicals and preservatives present in so many of today's new foods. Trying to process them creates additional toxicity in the body. It's a waste of precious energy.

Opting for whole natural foods is an important way to maintain your *Natural Energy.* This doesn't have to mean organic foods, although they're not bad. It simply means trying to avoid the foods that have had all the nutrients leached out of them and, whenever possible, to steer clear of foods full of preservatives and other unnatural ingredients.

Detox your body from overly processed foods for a week or two, and pay attention to how you feel. You'll not only feel lighter and more energetic, you'll soon find your cells craving the foods that are best for them.

Less *Really Is* More

If food converts to energy, does more food give you more energy? Quite the opposite. Eating too much food, whether at one meal or over the course of a day, puts tremendous stress on your cells and, as a consequence, on all your body systems. Those busy cells have to work even harder to deal with an overwhelming amount of food. They're like Lucy and Ethel on the chocolate-factory assembly line, desperately trying to stay ahead of all those bonbons.

If you're a Sign #3 who yearns for an afternoon nap, it could be because you're eating too heavy a lunch. A big lunch is fine if you have the time for a siesta, but if you have to go back to work, it can be disastrous. Instead of being alert and ready for action, you're drowsy and ready for sleep.

Get Off the Sugar Roller Coaster

Sugar might seem to be a food that provides high energy because it is burned rapidly by the body. It's already broken down into one of the three foodstuffs, right? In fact, the opposite is true. The quick rush we feel from sugar is short-lived. Sugar is immediately transformed into glucose as it enters the circulation and raises blood sugar levels quickly. That rush of energy and elimination of hunger feels good . . . for a while. But the brain reacts to this sugar abundance by telling the pancreas to produce insulin, the hormone that allows your body to use sugar. Remember, your cells think every meal will be the last. This insulin surge leaves you depleted of your energy. Now you'll eat anything in sight in order to quell that feeling. Unfortunately, the foods you crave contain those same sugars that got you into trouble in the first place. Refined sugars are particularly troublesome. They include candy, cakes, sodas, sugar-coated cereals, and white rice, white bread, and pastas made from white flours.

One of my patients is a forty-six-year-old attorney who works like a dog every day until she just can't think anymore. She seemed indestructible, until you realized that she kept going only by eating a high-sugar food every two or three hours. Her desk drawers were full of Twinkies. To no one's surprise, she got a virus which sent her to bed, and I seized the opportunity to give her a nutritional makeover. I prescribed 500 mg of carnitine and 30 mg of Co Q10, as well as 500 mg of magnesium, three times a day. She took a week off to sleep and to eat energy-enhancing foods. For this attorney, the evidence was in. She saw what a difference this made in just a week, and now she's eating better and exercising too. She's still working just as hard, but now she's helping, not hurting, herself.

Go for the Slow Burn

If you want a wood stove to heat your house all night, there's a simple technique. First, make sure you have a nice bed of hot, burning coals and then add as much new wood as the stove will hold. The new wood will take advantage of those hot coals to burn slowly all night.

Think of your body as an energy-producing wood stove and your cells as those hot coals. Eat the foods that release their nutrients slowly, which gives the cells time to turn the nutrients into the slow, steady flow of glucose that your brain and body need to function. Foods that burn slowly make you more energetic. The foods on the mitochondria-friendly food plan are those very slow burners, designed to give you a reliable supply of energy all day and all night long.

Fat Facts

If sugar is a food that can fool us, the problem with fat is even more complicated. We definitely need fat to survive, yet all the dietary advice tells us we must lower the amount of fat in our diets! No serious nutritional guidelines would advise us to eliminate it altogether. A diet without any fat results in dry skin, dull hair, a lack of energy, and a lack of taste in our food. It also forces the body to expend a lot of energy to make fatty acids from either stored fat or muscle. That's why people on extremely low-calorie and low- or no-fat diets feel drained and hungry all the time.

Fat is also important for maintenance of cell membranes and mitochondrial membranes, the headquarters of energy production.

The downside is that fat is—well—fattening. One gram of fat weighs in at 9 calories, whereas one gram of carbohydrate or protein is only 4 calories. What you don't burn gets stored away for future use.

Another reason not to gorge on fat is that a fat-laden meal will make you feel tired and sluggish. Your enzymes have to slog through a lot of useless calories in order to extract nutrients. If you feel stuffed after a high-fat meal, imagine how your cells feel.

So what's a body to do? Be sure to include some fat in your diet. But the secret is the type of fat you eat. All fats are not created equal. The saturated fats found in meats and in certain oils should be avoided because they clog your arteries and raise your cholesterol levels. Olive oil, on the other hand, is good for you, as evidenced by the low incidence of heart disease in the Mediterranean countries where olive oil is a mainstay of the diet.

Another important source of "good" fat is fatty fish like salmon, mackerel, albacore tuna, bass, anchovies, and sardines. These fish contain omega-3 oils which contribute to the proper functioning of the cell membranes and also help give skin a healthy glow.

If you don't eat a lot of fish but still want to obtain sufficient amounts of omega-3 oils, get some flaxseed at your health food store. You can use it whole or grind it up. Sprinkled on your morning cereal or mixed into yogurt, fruit juice, or vegetable juices, flaxseed will make an important contribution to your overall health and energy. Be sure to keep it refrigerated.

Do You Need to Lose Weight?

Any discussion of a new approach to eating always raises the question of weight management. This eating plan is not a weight loss diet. If you follow these guidelines, you will never feel hungry or deprived. That would defeat the very purpose of the mitochondria-friendly food plan.

However, eating for energy increases the odds that you'll find your weight management easier. Going on a low-calorie starvation diet is the *worst* thing you can do if you're looking for higher energy and lower numbers on the scale.

Most diets are based on the premise that when you put less food into your body, your body will break down its stored fat to access glucose for the brain and the other substances needed to support life. This is totally false. It is much easier for the body to break down its *proteins* before breaking down its *fat.* That sort of dieting has no effect on weight loss and a big effect on sapping your energy. It also breaks down precious muscle fiber.

So eat! Eat good food in appropriate quantities. Eat foods that are natural that the body has experience dealing with and using most efficiently. Avoid the obviously overprocessed, high-fat foods that your body can't use, and which it has to store as fat.

One thing you can count on with this food plan is that you won't gain weight. And if you need to lose a few pounds, experience will teach you how to modify the food plan to help you reduce your weight without reducing your energy level.

Watch Your Portions

When it comes to food portions, size matters. It matters even more when it comes to increasing your energy. Reasonable portions give your cells and enzymes the ability to work at a steady pace, whereas too much food puts stress on all the body systems engaged in extracting nutrients from the foods we eat.

But don't go crazy weighing and measuring everything you eat. That's an energy zapper if I ever saw one! Instead, try this method: Imagine what your meal would look like on a restaurant plate—a normal plate, not one of the giant-size platters that have become popular in some eateries in recent years. This will also help you present your food in a lovely manner, even to yourself, contributing to the overall pleasure of the meal. If you are an average-size person, go with the following portion guidelines. (Women should be on the lower end of the range and men on the upper.)

- MEAT: Four to six ounces, the size of a deck of cards.
- FISH: Six to eight ounces.
- TOFU: Four ounces. Tofu is usually sold in one-pound containers, so it's easy to cut a quarter of it.
- BEANS, PASTA, RICE AND OTHER LEGUMES AND GRAINS: About a half to three quarters of a cup.
- VEGETABLES: Restrict starchy veggies like corn, peas, lima beans, and squash to about half a cup. For leafy green vegetables and salad greens, eat as much as you like, but try to eat at least half a cup with lunch or dinner.
- OLIVE OIL: Though it's good for you, it is still a hundred percent fat calories, so try not to consume more than a tablespoon or so a day.
- FRUIT: A whole orange, apple, pear, or banana is one serving of fruit. Half a grapefruit, a quarter of a melon, a cup of berries, and half a cup of pineapple are good guidelines for these sweet fruits.
- JUICE: Six to eight ounces. Juice is high in fruit sugar and calories, so don't drink too much of it.

The Mitochondria-Friendly Food Plan

Now that we know how our cells work to turn food into energy, it makes all the sense in the world to give them those foods that are easily converted into *Natural Energy.* We know enough about what all foods are made of so that we can choose to help our cells by giving them foods with high concentrations of the good stuff they need to make energy. It works against our energy needs to torture our cells by giving them foods that are toxic to them or that make them work overtime.

By using this plan, you will select the foods and an eating schedule that maximize your cells' efficiency at extracting vital nutrients and leave them additional energy to run all your body systems smoothly.

A wonderful bonus is that many of the foods on the mitochondria-friendly food plan are also good for you in other ways. They contribute to keeping your cholesterol levels low and preventing your blood pressure from rising. They offer protection against free radical damage and cancer and can help you maintain strong bones. The increased energy you'll obtain from this food plan can inspire you to exercise more, which leads to even greater health benefits.

Put the right fuel into your body and you'll see the results in a smoother-running system. Eating for energy with the mitochondria-friendly food plan will transform your body from a sluggish machine that may seem over the hill into one capable of operating at the top of the mountain.

Learning to incorporate the principles of this high-energy, mitochondria-friendly food plan into your life will be easy. You don't have to worry about every meal. You don't have to weigh and measure your foods. You can certainly enjoy a varied diet at home or in restaurants.

Don't load yourself down with rules and daily requirements. Take a jigsaw-puzzle approach when considering your next day's food or even your next meal: look at what you've already eaten and fill in what you may have missed. If you haven't had a piece of fruit all day, make a crisp apple your afternoon snack or dessert. Did you skip a salad with last night's dinner? Have a big one with lunch today. Was lunch a lot of vegetables with a little pasta or bread? Put some protein on your dinner plate tonight.

SCHEDULE YOUR MEALS

When you eat is as important as *what* you eat. Although I don't want you to feel you have to be perfect, try to follow a schedule whenever possible. Our bodies run on a circadian rhythm in which hormones are released at specific times during a twenty-four-hour period. If we maintain an erratic eating schedule, we can disrupt our natural body rhythms.

When you wake up in the morning, your body expects food. If you wait too long to put food in your stomach, you are likely to succumb to a midmorning slump. I recommend that you eat a light breakfast, even if it's simply half a bagel and a slice of low-fat cheese on the run. If you really don't like breakfast, don't force yourself to eat it, but consider the fact that eating something in the morning is important to your energy level all day. Breakfast gives your metabolism a boost and your cells a chance to begin extracting nutrients while you begin your day's activities. You're also less likely to overeat at lunch if you've had a good breakfast. If you simply can't stomach even the thought of food right after you get up, eat a light midmorning snack as soon as you can.

Even if you eat breakfast, you may require a light snack by ten or eleven to keep you going until lunch, especially if you are a Sign #2 (waking up more tired than when you went to sleep).

Snacking is fine as long as it conforms to my recommended snacks, listed on page 152. And be sure it really is a snack and not another meal, which can actually make you feel tired, not to mention add unwanted pounds.

Lunch should be eaten at midday, not midafternoon. If you wait too long, not only are you likely to overeat, but you will also feel extremely wiped out. A late lunch typically leads to a late dinner, which will interfere with your ability to sleep. A light, high-energy midafternoon snack is fine, especially for Sign #3s who yearn for that midafternoon nap. If you need more energy for the afternoon, make your power lunch more powerful by concentrating on protein, keeping starch at a minimum, and keeping the whole meal a bit smaller. Eat very slowly to allow for optimum nutrient absorption, and do not drink alcohol. Skip dessert. We can learn a lesson from our Olympic athletes, who eat this way whenever they are in training.

I am an advocate of early dinners so that you have at least two hours to digest your food before going to sleep. I don't advise eating a snack before bedtime, because it can disrupt sleep. However, if you're hungry you can have a bedtime snack as long as you keep it light.

If you're a Sign #2 who needs to be sharper in the morning, start planning for that busy morning meeting or multitude of errands the night before with the proper dinner. Eat early, before seven, and eat protein—meat, fish, tofu—with plenty of good veggies. Don't drink liquor, and avoid dessert. Drink plenty of water throughout the evening.

LET'S START WITH MAGNESIUM

Magnesium jump-starts the production of enzymes that are essential for the production of ATP. It is also important for a wide variety of body functions—from building stronger bones to regulating body temperature, from regulating heartbeat to helping facilitate blood flow.

The bad news is that most of us are at least marginally deficient in magnesium. But here's the good news: There are plenty of tasty foods rich in magnesium. You don't have to eat all of them to keep your magnesium levels up; choose those that work for you and try to include them in your diet. I'm not going to tell you how much of any of these foods to eat. If you incorporate them into your diet, you will almost certainly get enough. And anyway, I think eating should be a pleasure, not a test! Eating a variety of the following foods regularly will have a dramatic effect on your energy levels:

- WHEAT BRAN: Sprinkle some onto your cereal or yogurt.
- WHOLE GRAINS: Eat brown rice instead of white rice, whole-grain breads and pastas, and muffins made of wheat instead of refined white flour. Explore the good tastes and easy preparation of grains like quinoa, barley, buckwheat, and millet.
- LEAFY GREEN VEGETABLES: Spinach, broccoli, broccoli rabe, collard greens, kale, and other greens are rich in magnesium as well as other nutrients, and can be prepared easily. Steam them or sauté with a little olive oil and garlic for the kind of high-energy delicacy our healthy Mediterranean friends have been enjoying for centuries.
- MILK: It doesn't have to be high-fat—skim or one percent milk has all the benefits without the extra calories. If you're a vegan or simply interested in some alternatives to animal products, soy milk and rice milk are good choices. They're available in more supermarkets all the time, and have many nutritional benefits beyond increasing your magnesium intake.

- MEAT: When you do choose to treat yourself to a meat meal, you'll get a bonus in magnesium.
- BANANAS, apricots, dry mustard, curry powder, nuts, and seeds are all rich in magnesium. All can be great ingredients in many dishes.

Just as some foods are rich in magnesium, to your overall health benefit, there are other foods which block maximum magnesium absorption and promote its loss through the urine. These foods—sugar, chocolate, soft drinks, highly processed meats, and excessively salty foods—are the suspects in a long list of other nutritional crimes, so try to minimize their use, or at least balance them with magnesium-rich choices whenever possible.

FIBER

If there is one type of food that can truly be called miraculous, it is fiber. Fiber is an indigestible food substance, found in plants, that is not digested or absorbed by the body. Fiber itself does not provide any calories or nutrients, but foods high in fiber are rich in vitamins and minerals.

In terms of conserving precious energy, fiber-rich foods are your cells' best friends. It takes smaller quantities of them to give you a full, satisfied feeling. They release all their benefits slowly, which allows the cells to extract nutrients with much less effort. Then these fiber-rich foods graciously leave the body with ease and efficiency.

Numerous studies have shown that a diet high in fiber can help prevent colon cancer by maintaining bowel regularity. Fiber helps lower cholesterol, and thus can reduce your risk of heart attack and stroke. Fiber-rich foods are a stabilizing influence on blood sugar levels. They are "slow burn" foods, raising blood sugar slowly and steadily to keep your energy level high all day.

Fit fiber into your diet with these tasty and satisfying foods:

* ALFALFA SPROUTS: High in fiber and low in cholesterol and fat, they make a tasty garnish on salads and sandwiches.
* APPLES: One medium-size unpeeled apple provides a whopping ten percent of the recommended daily total of fiber. Unlike sweeter fruits, apples help to regulate blood sugar levels.
* OTHER FRUITS: Most fruits are high in fiber. Enjoy fruit at every meal and between meals as well. Keep those fruits high in sugar—coconuts, pineapples, etc.—to a minimum and avoid fruit alone for breakfast, especially if you're prone to low blood sugar. (Sign 2's, take note!)
* BEANS: This great source of protein also provides plenty of fiber and works to prevent colon cancer while lowering blood cholesterol levels. Black beans make delicious soup, and are an excellent main course when combined with rice to create a perfect protein.
* BROCCOLI: Along with its cousins—cauliflower, cabbage, kale, collard greens, broccoli rabe, etc.—broccoli is a great source of fiber and of other essential vitamins and minerals that fight cancer.
* BROWN RICE, WILD RICE, AND OTHER WHOLE GRAINS: Brown rice contains about twice the fiber of white rice. Make brown rice a staple of your diet—along with barley, quinoa, millet, and buckwheat—and you'll experience a surge in energy and nutritional satisfaction. Try whole-wheat pasta as a great alternative to pasta made with refined white flours.
* CORN: That corn on the cob is good for you, but keep the butter to a minimum.

- LENTILS AND OTHER LEGUMES: Lentil salad is delicious. Mashed chickpeas combined with some garlic, lemon, and olive oil make a tasty hummus dip.
- OAT BRAN AND WHEAT BRAN: Mix into yogurt and add to your cereal for the best access to fiber you and your body can find.
- POPCORN: Your best choice as a snack food is dramatically lower in fat and sodium than potato chips, and a great source of fiber. Use a tiny amount of butter or margarine and sprinkle with a little Parmesan cheese—delicious.

HIGH-ENERGY MEALS

HIGH-ENERGY BREAKFASTS: Eating fruit alone raises your blood sugar quickly but drops it just as fast, and leaves you sluggish by midmorning. Have some whole-grain cereal—no added sugar or sugar coating—with low-fat milk, fruit, and nuts. Here's where it's good to add flaxseed, oat bran, or wheat bran. Try a tofu shake for breakfast. It's simple to make: Take a quarter of a cup of tofu or a cup of soy milk and mix it in a blender with a cup of fresh juice, half a cup of fresh fruit (banana or partially thawed berries are great), half a tablespoon of honey, and half a teaspoon of vanilla. Blend until smooth and drink up. It's delicious, and the energy you'll have all morning will amaze you. A cup of coffee in the morning is okay as long as you don't add sugar or sugar substitutes.

HIGH-ENERGY LUNCHES: Lunch can be the most important meal of your day—try not to skip it, no matter how busy you are. A good lunch refreshes you after your morning, provides a necessary break and rest in the middle of the day, and supplies your energy for the rest of the afternoon. Have some protein and a lot of fresh, colorful vegetables—grilled chicken or salmon with steamed veggies, for example. A salad with whole-grain bread is good, but add some protein—a little cheese or tofu, some grilled chicken or shrimp. Have a tuna salad sandwich on any bread except white. A baked potato stuffed with veggies and a little cheese is another wonderful one-dish meal. Minestrone soup is a great choice—vegetables, pasta, and beans combine for a nearly perfect nutritional boost. A main course of pasta may not always be a good lunch choice—it can make you feel tired in the afternoon, especially if you're prone to low blood sugar. If you can save the glass of wine for dinnertime, you'll be doing your energy level a favor, but if you decide to indulge, drink lots of water along with your wine of choice.

HIGH-ENERGY DINNERS: For most of us, dinner is the main meal, especially if we have families to feed. A leisurely dinner with warm conversation can be a wonderful reward to look forward to all day.

Create a dinner of two or three choices to give yourself the chance to savor a variety of foods you may not have had time for at lunch. You might start with a fresh salad composed of several different greens, some nice sliced cucumbers, and ripe tomatoes. Make a dressing of good olive oil and vinegar and flavor it with garlic and fresh chopped herbs or a little Dijon mustard. Move on to a cup of soup. Lentil or split pea soup is flavorful and nutritious, and even if you make it with a little bacon or ham, it's still relatively low in fat. Make your entrée a small portion of meat or fish with plenty of veggies and your choice of one complex carbohydrate—rice or pasta or bread or potato. Tofu and other soy products are a miraculous source of protein and other essential nutrients.

Beans and legumes such as lentils and split peas are also wonderfully high in protein and can be prepared in a myriad of ways. Combining beans with rice, preferably brown rice, results in a perfect protein—one that is low in fat and calories and high in the energy-providing foodstuffs our cells crave. Fruit for dessert is always a good choice, but don't have a cup of tea for at least half an hour after eating—it can inhibit the absorption of nutrients into your system.

HIGH-ENERGY SNACKS: Trying to fill our days with all the activities that we love can mean long hours between meals. Breakfast at six-thirty and lunch at twelve-thirty leaves too many hours during which we and our cells get tired and hungry. By ten in the morning or three in the afternoon, many people need a snack to keep them going. If you feel the need for a snack, go ahead, but stay away from candy bars, junk food, and soda. Opt instead for a handful of almonds or walnuts, or a small piece of low-fat white cheese (real cheese like Jarlsberg or Alpine Lace, not the artificially colored, processed kind!) with an apple or even a glutamate bar if your health food store carries them. Don't add a cup of coffee if you can help it—that caffeine will give you a burst of energy but you'll soon feel tired and depleted. Power bars, protein bars, and balance bars are also fine for an occasional snack. (Some of them are somewhat high in calories. If you need to lose weight, they may not be the best choice for you.) Flavored soy milk or soy shakes (which are sold in health food stores) are okay, too.

✵ A WORD ABOUT SALT—YES!

Some salty foods are actually good for you, in moderate amounts. Salt helps prevents dehydration by keeping fluids in your body. When you're sweating, it's particularly helpful to eat some salt to guard against fluid loss. Stay away from preserved meats—they are hard to digest and toxic to your liver. Do incorporate foods like chicken stock, miso soup, and nuts into your regular diet. In some but not all cases, salt may aggravate high blood pressure. If you have high blood pressure, check with your medical practitioner before adding salty foods to your diet.

A WORD ABOUT SODA—NO!

Despite the billions of dollars spent on advertising urging you to make your life better with a carbonated soft drink, the only one I can endorse is plain sparkling water. If I had my way, no one would drink any of the sugary, chemical-laden soft drinks so prevalent on the market. Diet soft drinks are no better.

BUT COFFEE IS OKAY

Don't panic—you don't have to give up that morning cup of coffee. There has been a lot of debate in recent years about whether coffee is a plus or a minus in your overall dietary plan. Does that first cup contribute to jump-starting the body's energy for the day ahead? Or does it set you up for a crash later on, which results in your drinking more coffee and remaining on a roller coaster of energy ups and downs all day?

The answers to these questions depend on the rest of your diet and personal health picture. Coffee isn't food. It is a signal for the brain to release certain stimulants into the circulation. This helps the body get the message to speed up. Now, if your body cells are well fed and have healthy, efficient mitochondria, a cup of coffee in the morning is like the turn of the key in the ignition. This is fine.

The problem with coffee and other caffeine-heavy products like soda arises when you are not on solid ground nutritionally or energywise. In this case, drinking coffee is like trying to push the body to respond when you have no fuel in the gas tank. Instead of working in concert with your well-tuned body, it increases stress and releases chemicals that can damage the heart and upset the body even more.

So if you like your coffee, take care of the rest of your nutrition with the mitochondria-friendly food plan and enjoy it like anything else—in moderation. If you would like to cut your caffeine consumption gradually, turn to herbal teas or a fresh glass of water for your between-meal refreshments.

INDULGE OCCASIONALLY

To go through life without ever enjoying a candy bar, a rich dessert, a glass of wine, or even a margarita is unrealistic and no fun. If you go to a big party and don't even touch the bowls of chips and fat-filled dip, the juicy barbecued hamburgers and creamy pasta salads, you'll feel unhappy and deprived. Sometimes a nice ice cream cone is just what the evening calls for. You can't run your whole life just for your mitochondria.

So allow yourself to savor some of these treats, and be willing to pay the price. You will feel a little sluggish the next day and your pants might be a little harder to button. Now that you know how your cells work with food, you may feel a little guilty for stressing them. Sometimes it's worth it, sometimes it's not. Sometimes you just *have* to do it. Just don't waste a lot of energy beating yourself up when you do indulge.

WHAT TO EAT FOR SPECIFIC ENERGY NEEDS

- IF YOU WAKE UP RAVENOUS: Don't eat fruit for breakfast, because it will raise your insulin level fast and drop your blood sugar level even faster. Have a little juice, but balance it with oatmeal or an omelette (made with one yolk and two egg whites) or multigrain and nut cereal or some tofu or soy milk. This will tide you over until lunch and keep your energy level up.

- IF YOU WANT TO PREPARE FOR MORE PHYSICAL ACTIVITY: Remember that quality is always more important than quantity. Eating extra food just slows you down as your cells spend more energy processing the food you put in. Go for protein and good carbohydrates, don't forget your Energy Pack, and have a banana to keep your potassium level up.
- IF YOU HAVE PMS OR ARE IN PERIMENOPAUSE: Increase your intake of foods rich in protein, calcium, and magnesium to help control symptoms. Keep up your exercise. Being sedentary only makes your symptoms worse.
- IF YOU WANT DINNER TO REV UP YOUR LIBIDO: Contrary to popular belief, alcohol doesn't help. Neither does eating a big meal. A relaxing, light repast of fruit, cheese, salad, or vegetables will get you—and keep you—in the mood for love.
- IF YOU HAVE TROUBLE FALLING AND/OR STAYING ASLEEP: Have an early dinner and don't drink. Alcohol may help you fall asleep, but it will also awaken you during the night.

Start Slowly

If you've been feeling sluggish, changing your food choices will make a big difference in your energy level. If some of the ideas on the mitochondria-friendly food plan seem too new and different, start slowly. You might consider simply adding a piece of fruit to your breakfast and lunch and more vegetables to your dinner. Try doing one thing such as eliminating soda or cutting down on coffee after breakfast. Buy some brown rice and flavor it with a little soy sauce.

After making just a few changes, assess how you're feeling. There's nothing more motivating than feeling better, both physically and mentally, as you see and feel the results of your new eating habits. Gradually incorporate a few more of the suggestions in this chapter, like eating dinner earlier or cutting out a few trips to the fast-food emporiums.

The new energy you gain from the mitochondria-friendly food plan will inspire you to begin to make other changes in your life to restore and maintain your energy system. The lifestyle suggestions in Chapter 8—"Revive!"—will be easier to implement when your body and cells are fully nourished and you're feeling better and more optimistic.

Chapter Eight

REVIVE!

There was a time when I believed that the definition of efficiency was packing as much into my days (and nights) as possible. I was a woman in perpetual motion. I felt honor-bound to see every patient who called the office right away, whether or not it was an emergency. I felt compelled to accept every social invitation that came my way. I wouldn't want you to think that I let my family responsibilities slide. In fact, they came first. I made it my business to be there whenever and wherever my kids needed me . . . whether they wanted me there or not! On those rare occasions when I had an hour or two of downtime I felt guilty, and I would look for ways to fill it up.

Then my energy system crashed. I was forced to slow down

and take stock. For the first time, I could see that my "She just keeps on going and going" approach to life was counterproductive. Contrary to my belief that the more you do, the more you get done, I now saw that the stress to which I was subjecting myself was sapping my energy and making me *inefficient.*

I also realized that my situation was not unique and I was not alone. I could see my problem mirrored in hundreds of my patients who themselves were caught in the throes of the energy crisis and whose lifestyles were feeding into their own exhaustion. I observed that we tend to make the same mistakes time and time again. We subject ourselves to energy zappers that not only leave us physically depleted but drain us of our health and well-being. I know how it feels when your energy system is failing, and I know how to pull out of this physical and psychic nosedive. In fact, I'm living proof that it can be done!

I think my strategy works because I don't rely on a single tactic. I attack the energy crisis in three separate but mutually dependent ways. The Energy Pack will replenish and recharge your aging mitochondria. The mitochondria-friendly food plan will help you eliminate the poor eating habits that drain your natural energy. Now, the strategy I call "Revive" will show you how to make critical lifestyle changes that will boost your energy level and heighten your spirits.

The point of "Revive" is that I don't want you to squander your *Natural Energy* on energy zappers that add nothing of value to your life or the lives of those around you. I want you to save your energy so that you have it to spend on all of the activities, challenges, and interests that truly make your life worth living.

Here's how you can do it. Over many years of practice, I've learned the four key steps for reviving your life:

1. De-stressing
2. Engaging

3. Moving
4. Sleeping

Sounds simple? It is. Yet incorporating these four basic steps into your lifestyle will help you look forward to the days, weeks, and years ahead with renewed zest and vigor.

Keep Track of Your Progress

I recommend that you keep a diary for your first three weeks on the *Natural Energy* program and for as long as you like after that. It's a handy device to help you keep track of your mood and activities. Your diary needn't be a separate book—you can just use your usual daily planner or appointment book. Jot down notes about when and what you eat, the times you exercise, the hours you sleep, and most important, what kind of mood you find yourself in. Write about your feelings when you want to. This will enable you (or force you!) to take the time to look at what you're doing, how you're spending your time, what your priorities are, and where there is room for change.

When you feel really good and energized, look back at the last few days and you'll discover why. Those healthy meals you ate at appropriate times, and the long walk you took this morning along with the stretching you did two days ago, will explain the good mood and great energy you're experiencing. On the other hand, if you're sluggish, look at your diary and figure out the explanation. Your actions and your mood will usually correlate with each other. Use what your mind thinks and feels, as well as how your body is reacting, to help you make decisions and choices each day.

You know, for instance, that both sleep and exercise are good for you. But there will be times when you'll have to choose between them. Check your diary. If you've exercised two or three times this week, opt for some extra sleep. On the other hand, if a week has gone by and you haven't even taken a short walk, the exercise choice may be worth pushing yourself for.

Your diary can be an important tool in helping you help yourself in specific ways without beating yourself up because you haven't done it all.

De-stress!

There is no greater energy zapper than unrelenting stress. The antidote is simple: De-stress!

I know what you're saying. "Easier said than done." That may be true, but only up to a point. Life is by definition stress. Everything we do stresses our systems in one way or another. Whether it is mental stress with family matters; personal issues; work-related challenges; or physical stress in terms of eating, rest, and exercise patterns, the result is wear and tear on the whole system. This wear and tear can't be eliminated, but it can be minimized, delayed, or at least modulated and balanced so that it doesn't zap your energy.

First, identify the stresses to which you are particularly vulnerable and figure out if you can do anything about them. When you are feeling pressured, get your diary and write down the events that precipitated that feeling. You are very likely to notice a pattern. For example, if you're a Sign #1 (your mind says "now" when your body says "later"), it will soon become apparent that you are making unrealistic demands on yourself. Take this as a signal that it's time to slow down, at least for now.

Let's say you're stressed out every time you have to say no to your children because of demands at work. It's possible to take steps to find more time—by rearranging schedules, by setting aside two evenings a week for activities you can share, by taking your children with you on a business trip now and then. It may not always work out perfectly, but the fact that you've taken action and can spend even a little more time with your children takes a big load off your mind.

But suppose your family stress is something you can't do anything about. Perhaps you're part of a blended family and your partner's daughter isn't ready to accept you yet. You can honestly say you've tried everything you could, but it's just not working. Now's the time to take an approach to a stressful situation you may not have considered before:

PUT IT IN AN IMAGINARY DRAWER AND SHUT IT!

For many people, this might seem like unusual advice, but I'm not talking about sticking your head in the sand.

Take Fran. At twenty-nine, she's a beautiful, successful executive who was in and out of my office for a couple of months with numerous complaints. She had headaches, chest pains, problems breathing, and a myriad of other physical symptoms. I examined her but each time found nothing wrong. Finally, I decided we had to address the real story. Fran had just taken on a new job and bought her first apartment. Stressful life turning points, to be sure, but her real problem was that she was in a dead-end relationship with a college sweetheart that she knew would never go any farther, and she didn't know how to end it. Bingo! Now I understood why she wasn't sleeping. And I saw why she kept coming in with so many symptoms. Her body was literally crying out for help.

I told Fran that she was worrying about too many things at once. Instead of telling her to stop worrying altogether, I advised her to focus her concerns on the one aspect of her life that was vital for her future—her job. At the same time, I advised her to stop fretting over her relationship. Since she knew it was going nowhere, if she chose to continue it she should simply accept the situation the way it was. After she put her anxiety over the relationship into a figurative drawer, something interesting happened. Now that she was no longer fixating on it day in and day out, within a few weeks she was ready to let go of it. She was also doing great at work!

That one piece of advice—stop fixating on a problem that you can't fix—made more difference to Fran than all the medical treatment in the world. Her symptoms have vanished, and she is happier than she has ever been in her life. And yes, she met a new man at work.

We are very accustomed to working out every problem, struggling with it until we find a solution. But there are some problems without solutions, at least for a while, and rather than continuing to stress yourself when there's no answer in sight, give yourself a break. Take the energy you're wasting on something you can't change and use it for the things you *can* do something about.

✹ PAUSE, EVEN FOR A SECOND.

It dawned on me recently while playing tennis that the difference between an average and a good tennis player is a split-second delay. If you can stop and think as the ball is coming toward you, you will play better. Take this concept and apply it to life. Before reacting to a stressful situation, take a split-second delay and figure out why you feel this way and what you want to do about it. You'll find yourself making a more intelligent choice. The split-second concept really means being aware, becoming aware of yourself. Don't react blindly. Don't eat something until you ask yourself if you really want it. Think for a moment before every action you take. This doesn't mean you should stop and intellectualize over everything you do, but do stop for a second and make sure you're doing what you want to do.

STOP FEELING STUCK IN THE PAST.

Feeling like a has-been can be tremendously stressful. So you used to be a college athlete but now you can barely throw a ball. Your former buns of steel are now soft biscuits. Though you never had a problem staying out until two a.m. and then wowing your colleagues at an eight-thirty breakfast meeting, these days if you're not asleep by midnight, you're useless all the next morning. Well, so what?

Let it go!

If you believe your best days are behind you, then that will be true. Instead, take a look at what you've got going for you now and deal with how you can make the present even better than the past. Use your new *Natural Energy* for today instead of wasting it on memories of yesterday.

I've got news for you. I'm not the woman I used to be . . . I'm better!

TAKE ONE STEP AT A TIME.

If you're often overwhelmed by all the demands on your life, including your own demands on yourself, you may feel that no matter how hard you try, you're going to fail at something. That's understandable.

The problems come when that feeling keeps you from moving ahead on *any* of your activities or goals.

One way to cope is to sit down and make a list of everything you have to do. The list will likely be a combination of tasks (cleaning the house, paying the bills) and goals (getting out for a walk or going to sleep early no matter what else is going on). Then give yourself a boost in the right direction by doing one or two of the easiest tasks on the list. By moving ahead, at least slowly, you'll break that pattern of powerlessness and show yourself that you can do what you set your mind to.

✳ HAVE REASONABLE EXPECTATIONS.

If you wake up in the morning and have only an hour to get to work, realize that you can't weed the garden or stop at the gym. You'll have to do those things another time. If you don't get home until nine p.m., don't even try to plan a candlelight dinner with your husband. Save it for Saturday night. If you have only half an hour to exercise, do so for twenty minutes. Ten extra minutes in the shower will move you on to your next activity without being rushed and frantic. Be realistic about what you can and cannot do. In other words, give yourself a break. Guilt over projects undone or unrealistic goals unmet is a major energy waster!

✳ SET ASIDE TIME TO WORRY.

Some people fixate on problems and worry about them all the time. Worry by its very nature is a true energy zapper: fearing that something might or might not happen in the future is utterly useless. It doesn't solve whatever problem you're focused on and it fritters away your energy. Over the years, several people have shared the following strategy for those who are prone to worrying: Rather than worry all the time, set aside a certain amount of time, perhaps half an hour every other day (you can even schedule it in your diary), and worry all you want about whatever it is. Then be done with it. Don't let it invade your every thought and waste your precious energy.

CLEAR UP MISUNDERSTANDINGS.

Do you find yourself in the same discussions and arguments over and over again with certain people in your life? Do you come away from many conversations feeling you weren't heard and you really don't understand what the other person was talking about? Poor communication skills are at the root of this problem. Often when people talk, they're playing verbal Ping-Pong. They don't talk *with* but *at* each other. They don't wait for the other person to finish the sentence; they're too busy preparing their answer. Try to slow down and really listen to what someone is saying to you. Take a moment to identify, at least partially, with that point of view, even if you don't agree with it. Ask a question to make sure you understand and then assure the person that you do see what he or she is saying. Take another moment or two to answer, and choose your words more carefully so that you make your point well. You'll save a lot of energy and improve your relationships as well.

GET ORGANIZED ALREADY!

Can't find your keys? Running late on paying your bills even though you have money in your checking account? Forgetting to buy basic groceries, so that cooking a simple meal becomes a big chore? These are big energy wasters, and disorganization is the culprit. If all of this bothers you and adds to your stress and exhaustion, try to figure out where you might make some changes in the ways you do life's business. Hang a hook for your keys by the door. Set aside one hour a week to look at the mail (which you can just throw into a basket each day) and pay the bills. Devote one Saturday morning a month to stocking your cabinets with the things you need. Investing in your future means more than contributing to your IRA. A little planning and organization will take some energy now, but it can save you lots of energy in the future.

GET A NEW ATTITUDE.

The most important de-stressor may be simply changing your own outlook. Take the attitude that life is good and, to quote a recent movie, that this is as good as it gets. Many of us believe life is about striving toward a future goal, but I think that can cause us to miss the present.

Rather than always trying to improve things for the future or blaming yourself and others for the past, learn to treat yourself kindly right now, in this moment. I'm not talking about pampering yourself with a massage or buying yourself more possessions, although an occasional indulgence can be restorative. I'm talking about giving yourself permission to be yourself, to be who you are. Give yourself permission to age but not to give in. Bless your own efforts to address these issues and feel good about yourself. If you're forty-five, don't fret over the fact that you're not twenty-five—be the best forty-five you can be. And don't use someone else's standards; define yourself as unique.

My patient Joe, a sixty-five-year-old former advertising executive, had recently embarked on a second marriage with Karen, a retired teacher. After forty years in the classroom, which had earned her a wonderful pension, Karen now wanted to use some of her hard-earned savings to treat herself and Joe to a six-month honeymoon traveling around the world. In his younger years, Joe would have leapt at the chance to travel. Now, he confessed that he felt tired and was worried about being too old. He was a true Sign #6—everything was going well in his life, yet he was stressed out and blue. On Medicare and a card-carrying member of the AARP, Joe felt his life might soon be over and he should be slowing down. A thorough checkup indicated that everything was fine, at least physically. A regimen of 500 mg of carnitine and 60 mg of Co Q10 daily helped to break the cycle of fatigue, and three months later he was a new man. He has started an exercise program, eats a healthier diet, and has lost ten pounds. Last I heard from Joe and Karen, they were somewhere in Tuscany taking a cooking class.

Along with working on your own attitude, taking responsibility for yourself will also give you more energy. If you feel bad, ask yourself why. Did you have an argument with someone or a disappointment that something didn't turn out as you'd hoped? Is it worth ruining your day over? Did you gain five pounds in three days, and are you angry about it? Look at your diary and see what you ate and whether you exercised. The answer will be there. So take responsibility for it, and then take charge of making the changes.

Engage!

Lack of stimulation is a major energy zapper. It can dull the mind and the body, and leave us depleted of the desire to even do anything. In fact, one of the main causes of Sign #4 (brain fog) is lack of meaningful mental activity.

The key is to prevent boredom from setting in. Many people think boredom is not having anything to do, but I think boredom is doing something you don't like, something that offers no stimulation to your mind. If you're just sitting and staring at TV and not interacting with others, you're going to be bored.

New people, new ideas, and interesting activities that teach you something you didn't know before, are never boring. I find myself bored if I'm not expanding my mind with a new project or some other stimulating new activity. I've also noticed that when I get bored, I often get sick. It's almost as though when the mind has nothing to do, it plays mischief with the body to create a form of diversion.

Boredom can also be a place to hide from your problems. The mind knows that if you address the reality of your situation, you'll have to change it, and that can be hard. So we often use the pretense of boredom to protect ourselves from matters we'd rather not face.

When you're alone, don't just sit there—read, write letters, do a crossword puzzle, interact with others by telephone or E-mail, or venture into a chat room on the Internet. Stay in touch with friends. Build up your contact with the outside world—through social events, church or synagogue membership, involvement with family and friends, and other "energizing" activities.

If, as you scan your diary, you notice that you go for days without meaningful human interaction, find a way to get back in touch with family or friends.

One of my favorite success stories is that of Andrew, a sixty-eight-year-old former construction worker and sports enthusiast who was fast becoming a total recluse, and thought of every reason in the book to stay that way. Andrew spent his days channel-surfing. I was alarmed because Andrew's arthritic joints were so stiff that he could barely move, and unless he began some type of physical activity, they would only get worse. He also seemed to be in a state of perpetual brain fog. I advised Andrew to go for a short walk each day, but he said he couldn't. Why? His joints hurt too much. In an attempt to get him to leave the house on a regular basis, I asked him if he ever went to church. He replied that he would like to but couldn't. Why? The service conflicted with his favorite Sunday morning news show! I saw my opening and ran for it. I suggested that Andrew tape the show. Out of excuses, he reluctantly followed my advice and went to church. The minister must have heard my prayers because he cornered Andrew at the end of the service and asked him to help supervise a construction project at the church. Within a few weeks, Andrew was back doing carpentry, and feeling so energized that he was actually able to walk the two miles to church daily. It took just a little bit of engagement to get him on the road to having a life again.

Move!

Everywhere you look, you are given the message that you have to exercise. Exercise is vital for cardiovascular fitness, lower blood pressure and cholesterol levels, better muscle tone, stronger bones, and flexible joints. Exercise clears your mind, releases positive hormones like beta endorphins, and helps you sleep better. The increased blood circulation you get from exercise gives your skin a healthier glow. No question about it, exercise will make you function, feel, and look better.

But here's the most important reason to exercise, in my view: It takes energy to make energy. If you don't move, you won't have the energy you want. When you get moving—even with a few stretches or a short walk—you generate the kind of pure, natural energy your body thrives on. Think about children. Do they always move because they have so much energy? Or do they have so much energy because they're always on the move?

The human body is designed to move. The muscles of the lower body—the hamstrings that run from the back of the knee to the bottom of the hip and the gluteus muscles around the hip, tailbone, and thighbone—are the largest muscles in the body and the ones that propel every step we take. When we spend most of our time sitting on them, they lose their elasticity and become weak, flabby, and no longer able to do their jobs. You may not care if you ever run a marathon, but you will miss out on a lot if you enter your later decades unable to walk down the street.

Many of the patients who come to me in the throes of an energy crisis object at first when I recommend that they exercise. How can they exercise, they say, when they don't even have the energy to get through the day?

The first thing I do is get them started on the Energy Pack. After a few days or weeks on carnitine and Co Q10, their natural energy is increased, and pretty soon they *want* to move.

Maintaining a regular schedule of exercise doesn't have to become a burden in your life. As important as exercise is, there are different ways of approaching it that will make sense for your individual preferences. If time is an issue, be assured that you don't have to devote endless hours to exercise to get positive results. In fact, some studies suggest that even a few short bouts of exercise daily—as little as ten minutes at a time—can have a profoundly positive effect on your mental and physical health. A short walk in the morning, a quick session on a stationary bicycle at home or at the gym during lunch, and even ten minutes of weight training in the evening are simple ways to get yourself moving. Fifty crunches on the floor at your office can be a great help for those of you with afternoon slumps.

Of course, the best approach is to do something that you'll enjoy. I personally hate organized exercise. I avoid gyms, but love tennis, and so I make it a point to play at least three times a week after work or on the weekends. In the winter, I walk on my treadmill while watching TV. If you've always loved swimming, make two or three half-hour sessions a week your whole approach to aerobic conditioning. Whatever activity or sport gets your heart beating faster—tennis, basketball, racquetball, running, brisk walking—you're more likely to stay with it if it's something you enjoy to begin with. Complement your aerobic workout with a few sessions a week of stretching and toning, perhaps some yoga or any kind of basic exercise class that appeals to you. Try working out with weights occasionally. It's great for your bones.

If you've been a couch potato for awhile, start slowly. Remember that any kind of moving is good. You can sit in your chair and stretch. Do a few stretches before you even get out of bed in the morning. Walk whenever and wherever you can.

Whatever you choose, don't let exercise add stress to your life. There will be times when you just can't get to it, and that's okay. There may be times when your need to sleep and rest outweighs your need to exercise. Be sensitive to how your body feels and give it what it seems to be asking for.

Sign #7s (your sex life is stuck in low gear) take note. Several studies have documented that people who exercise have sex more regularly than people who don't! There's another reason to get moving.

Sleep!

We rarely forget to eat. Why is it that we forget to sleep?

When I was growing up, my mother used to say, "You can never make up for lost sleep." She was right. Your body needs sleep—good, deep, uninterrupted sleep—as much as it needs the right foods. You need between seven and nine hours of sleep every night. You might get by on as little as six, but if you find yourself cranky, unfocused, and sluggish, increasing your hours of sleep will make all the difference in the world. As you learn to take better care of your body, and especially when you're trying to recover from one of the Seven Signs of the Energy Crisis, pay careful attention to how much sleep you get.

Sleep is critical to maintaining your energy system. Burning the candle at both ends will soon leave you with no candle at all. Some people think that when they're sleeping, they aren't accomplishing anything. Quite the opposite—the rebuilding of your energy system that occurs during sleep is a huge and important accomplishment indeed.

Here are some tips on how to make good sleep a priority in your life. Sign #2s (waking up more tired than when you went to sleep), pay close attention!

Clear your bedroom of everything that doesn't belong there—work papers, for instance. A calm and inviting atmosphere will add to the quality of your sleep.

Try not to just pass out on the couch in front of the TV, or leave the TV on in your bedroom all night. The light and sound of the television will continue to enter your brain and the quality of your sleep will be seriously diminished.

Look at how you're eating. If you're eating dinner too late, try to eat earlier. If you're eating heavy food, try to lighten up the menu. Going to bed lighter will definitely help you sleep better.

Don't exercise too close to your bedtime. Though it may make you fall asleep quickly, you're very likely to wake up during the night. But if you're exercising before going to sleep and you sleep better on those nights, keep at it. If instead you're falling into a deep sleep after exercise but then waking up frequently, try to exercise earlier in the day.

When you can't sleep, don't get worked up about it. Find the technique that works best to get you back to sleep. Clear your mind, notice your breathing, meditate if you like. All that calm and quiet just might put you back to sleep.

A Little at a Time

This advice may seem, at first, like a lot to think about incorporating into your life. But if you give it a chance, a little at a time, you'll find a whole new world of energy and excitement awaiting you.

Be nice to yourself. Your needs are yours alone. Taking the time to meet and understand yourself will make you kinder to yourself and a better member of the human community.

If you are tired, sleep. Things will wait.

When you are sad, cry. You'll feel better.

When you're happy, enjoy it. You'll make others happy too.

If you've been a couch potato and don't want to be one anymore, consider this: Why disregard the world and live in your own cocoon? Why would you deprive the world of what you have to offer and, more significantly, deprive yourself of what the world has to offer you?

Take a few weeks to make the changes I prescribe in this book and then get out there. Start your day by moving. Spend ten minutes in the shower, thinking about all the opportunities the day can present. Be happy you're alive. Think of someone you'd like to make happy and plan to connect with that person—today! Figure out at least one good thing about your job, even if it's a job you don't like. Look in the mirror and see a good person. Go shopping, go out, become part of the group spirit. Do something for someone else—volunteer to serve Thanksgiving dinner at a local homeless shelter, or help run the charity bazaar at your church or synagogue. Discover the *Natural Energy* and life that are yours for the asking.

Chapter Nine

SPECIAL BOOSTS FOR SPECIAL NEEDS

There are times in life when our energy systems are under unusual stress. Whether we're confronted with seasonal or chronic annoyances such as allergies, a potentially serious illness such as diabetes, or a normal life passage such as menopause, our bodies need an extra boost at these times.

The focus of conventional medicine is on treating each ailment or disease symptom by symptom. To my way of thinking, this only accomplishes half the job. The *Natural Energy* program bolsters the entire body so that it can better cope with whatever challenges come its way. My goal is not only to help people through a current crisis but also to fortify them so they can better handle the next.

This chapter will help you overcome those times when your energy stores may run down, whether due to illness, unusual circumstances, or a normal life event.

To make it easier to find the information you need, Chapter 9 is divided into two parts. The first offers tips on how to weather life passages from childhood through the later decades. The second reviews the treatment and prevention of common chronic annoyances and illnesses.

You may find the information in this chapter useful even if you don't have a particular problem, because the steps you take to cure an illness are often the same ones you need to take to prevent it.

Life Passages

CHILDHOOD

Children are typically in a state of perpetual motion. Their energy systems are in high gear, but the fact that they can seemingly run on empty is deceptive. We often forget that children are living on borrowed time—what they can get away with today invariably catches up with them tomorrow.

The seeds of heart disease, depression, diabetes, and obesity (to name just a few ailments that afflict adults) are sown in childhood. I am convinced that these diseases often could be eradicated if parents would teach their children to simply pay attention to their bodies.

Teach them to eat at regular times when they are hungry, not because they are bored or upset.

Keep the right foods in the house. I am astonished at the number of parents who are shocked that their children are overweight, yet stock their pantries with soda, cookies, and chips. In these homes, the main activity is invariably watching TV or sitting in front of a computer screen for endless hours.

Help your children develop the lifetime habits that will prevent an energy crisis from ever appearing. From an early age, teach them to eat the right balance of the different food groups. Teach them that a wide variety of colors on their plates means that they are getting a wide variety of vitamins. Even small children can understand that a meal should consist of protein, carbohydrates, vegetables and fresh fruits, not chocolate-covered cereal or processed luncheon meat on white bread.

I don't mean going to extremes and denying your children an occasional sweet or order of french fries, or making them feel weird at school because they're the only ones with tofu sandwiches in their lunch boxes. Even fast-food restaurants aren't off limits if children are taught how to order sensibly. However, feeding your child a steady diet of junk will take its toll.

Get your children moving! Don't allow them to spend days and days without getting exercise. It's bad for both their bodies and their self-image.

Teach your children how to cope with stress in a constructive way. They should be taught to rest when they are exhausted, and how to identify the signs of exhaustion. And it's important to teach them how to articulate their concerns when they are under emotional duress. Helping children to choose the right words to describe their feelings is key to rearing them to be happy, healthy adults. Too often, adults minimize the concerns of children, and make them feel silly and insignificant. Although we don't want to blow things out of proportion, it is important to acknowledge that children are entitled to feel upset. As adults, once we understand their concerns, we can direct their attention to problem-solving, and not teach them to conceal their feelings.

ADOLESCENCE

The working definition of adolescence is burning the candle at both ends. I am convinced that the only people who are up at night watching the *Late, Late Show* are under the age of eighteen! They don't have the sense to turn off the TV and go to sleep.

Granted, teenagers have strong energy systems, but they are not indestructible. I have seen too many teenagers ignore the signals their bodies were sending them and end up doing poorly in school, sick, and depressed. I've seen innumerable parents just as blind to their kids' energy crises.

Teenagers are typically overwhelmed by their bodies. When hormones begin to kick in, not only are they inundated with new urges and desires, but they begin to see the world quite differently. Children want to be safe and protected, but teenagers want to push to the limit, and tax themselves and their parents.

Parents respond by implementing curfews or harsh punishments, but that is not the most effective approach. Adult supervision is important, but the fact is, teenagers need to be armed with information to make their own choices. We can't be there for them every second, even if we wanted to. As a mother of two daughters—one teenager and one young adult—I have found that teaching a teenager to be self-aware is a far better approach. It is a continuation of everything I believe we should be doing from childhood on, but in greater detail at this crucial time.

Teenagers need to be taught how to put balance and perspective into their lives. They need to be shown how they can do teenage things without destroying their bodies. For example, when they're going to be up until three partying on Friday, they need to plan on catching up on their sleep Saturday night, not partying again until the wee hours of the morning. If they run themselves ragged, they will eventually get sick. If they get sick, no more parties. It sounds so simple, yet so many teenagers don't get it. Come to think of it, many adults don't get it either.

So let's try to promote the ethic of self-awareness in our kids. If we can teach them early to pay attention to what their bodies are saying, they will grow into more resilient, healthier adults less likely to suffer from an energy system meltdown.

POSTPARTUM DEPRESSION

Postpartum depression has a lot in common with adolescence. Like a teenager, a new mother must contend with wildly fluctuating hormone levels and a sense of disconnection between her body and her mind. Even though she has given birth, she is still carrying extra weight and is definitely not feeling like herself. How could she? Her breasts are still swollen, her body may still be sore from the delivery, and she's exhausted. Unlike a teenager who can sleep it off, a new mother gets very little of that luxury. For all these reasons, I don't think there's a new mother who has not experienced some form of postpartum blues.

It is worse for women today than ever before. Very often, women in the throes of postpartum depression are now also feeling pressured to return to their jobs. In this era of drive-through deliveries, society also expects women to bounce back from childbirth instantaneously. It wasn't always this way. There was a time when women were encouraged to stay in the hospital for up to a week after having a baby to regain their strength and learn how to care for their newborn. No more. Today, women are expected to return home in forty-eight hours, be back in full swing in a week, and down to a size 6 in a month.

What is not considered is the fact that after a woman gives birth, her energy system, not to mention every cell in her body, is in shock. For nine months her body has focused exclusively on nourishing and protecting the growing fetus. After the baby is delivered, her body needs time to recoup. But our fast-paced society doesn't accommodate new mothers.

A new mother must ignore the negative signals from outside and tune in to what her body needs. For at least three months, the focus of her life should be taking the best care possible of her baby and herself.

Don't put yourself under unnecessary pressure such as trying to lose weight overnight. Rest assured, the weight will come off naturally over time. Starving yourself, taking diet pills, or exercising beyond your endurance is foolish. In the long run, it will wear you down.

Take your Energy Pack with an additional supplement, DHA—a form of omega-3 fatty acids that can help prevent depression. Because of our modern diet, new mothers are often deficient in DHA, and some researchers believe that this is a contributing factor to postpartum depression. Take three 250 mg capsules daily. Since DHA is added to infant formula in many parts of the world, it is safe to use even if you are breastfeeding. DHA is available in health food stores.

Get enough sleep. Your sanity depends on it. When your baby sleeps, you should sleep too. The housekeeping will wait. If you have other young children who need supervision, try to get some baby-sitting relief during the day so that you can take a nap. I have found that exhaustion in new mothers can trigger destructive eating habits that can continue for decades. I wonder how many women with Sign #5 first began overeating after the birth of their children.

If you work outside the home, give yourself as much time as you can before you return to work. I understand that there are bills to pay and careers to consider, but you need time to recharge. It is not a luxury but rather a medical necessity.

PERIMENOPAUSE AND MENOPAUSE

Perimenopause is the beginning of the hormonal changes that eventually lead to menopause, the cessation of menstruation. The average age for menopause in the United States is fifty-one, but perimenopause can begin a full decade earlier.

Insomnia is often the first sign of perimenopause. Typically, a forty-something woman will complain that she is exhausted because she wakes up every hour. When I explain that this is due to a decline in estrogen, invariably she responds with a stunned, "I'm way too young for that."

Many women still view menopause as a single life event that hurls them into old age. That may have been true at the beginning of the twentieth century, when the average life expectancy was around fifty, but it is no longer the case. Today, menopause is a natural transition that moves women from one stage of life to another. It is not a signal of the beginning of the end; rather, postmenopausal women today can look forward to living three, four, or even five decades longer.

Of course, you want to live out those years in a strong, healthy body. Perimenopause provides the perfect opportunity to begin to prepare your body for the next half of life. Now is the time to evaluate your lifestyle and make necessary corrections. Getting yourself on a regular exercise routine and following healthy eating habits can prevent numerous medical problems that can rob you of your ability to live a full, vital life down the road. Start to include calcium-rich, cancer-fighting foods such as soy milk, tofu, and yogurt in your diet. Eat more fiber-rich fruits and vegetables. If you smoke, stop. Smoking is a shortcut to heart disease, lung cancer, and osteoporosis.

Take your Energy Pack to support your energy system. Add 400 mg of magnesium and 1000 mg of calcium to protect your heart and bones.

If you are experiencing any unpleasant symptoms, such as hot flashes, night sweats, or irritability, hormone replacement therapy can bring fast relief. There is a variety of hormone replacement regimens, ranging from synthetic hormones in blister packs from the giant pharmaceutical houses, to natural hormones compounded in special pharmacies and even to natural plant-based products sold over the counter in health food stores. You probably need testosterone too. You need to work with a knowledgeable physician who can help you select the regimen that's right for you. If I were writing this book five years ago, your chances of finding a doctor who understood the needs of menopausal women would be slim. Fortunately, today many more doctors are specializing in this field. Be sure to find a doctor who doesn't just hand out a prescription, but fully discusses your options with you.

My final word of advice is, don't waste the precious perimenopausal years with your head in the sand, pretending that your body isn't changing. Use them as an opportunity to make positive, life-affirming changes that will enable you to get the most out of life at any stage.

Chronic Annoyances and Illnesses

ALLERGIES

Allergies are heightened sensitivities to substances such as plants, foods, animal dander, chemicals like laundry detergents, or even medicines. An allergy is caused by an overreaction in the immune system. When the body encounters the offending substance, it reacts as if it were battling an infection. In response, chemicals known as histamines are released into the bloodstream, and this causes the itchy nose and throat, watery eyes, skin rashes, and other symptoms typical of allergies. Why some people get allergies while others don't is still a mystery. Genetics plays a role, but it is not the whole story. When you're tired, run-down, stressed out, or ill, you are more likely to suffer from allergies than when you are rested and well. A body sapped of energy does not have the stamina to function properly. The weakest system—in the case of allergies, the immune system—will falter.

Allergy testing only tells half the story: Just because you test positive to cat dander doesn't mean you will have an allergic reaction to it. It is also likely that you will develop allergies at certain times in your life and outgrow them at others.

We are so used to running for a pill or an allergy shot that we never give ourselves a chance to evaluate the cause of the allergy. My advice is, when allergy symptoms strike, take these simple steps first:

Purify: Clean out your entire system. For one week, drink a gallon of water a day. Steer clear of milk and milk products because these foods can increase mucus secretions. Avoid herbal teas that contain ingredients such as chamomile and other botanicals.

Ask Yourself What's New: What have you done within the past forty-eight hours that could have triggered the allergic attack? Are you using a different laundry detergent? Are you renovating your house? Are you pulling up old carpets, which could cause dust and mold to circulate in the air? Have you started your spring planting? Is the pollen count high?

Practice Avoidance: If you can pinpoint the cause of the problem, the solution is obvious: Avoid it! If something in your house is causing your problem, get rid of it. If it's the height of the allergy season, stay indoors with the air-conditioning on, and so on.

Take Your Energy Pack Plus: Take the Energy Pack with 1000 mg of vitamin C daily.

If these simple steps don't work and you don't feel better within a week, then it's time to see your doctor for treatment or further testing. I am not a big fan of over-the-counter antihistamines. They either make you drowsy or contain a stimulant that can make you edgy and interfere with sleep. Some of the new prescription medications, however, are excellent. Before using any allergy product, talk to your doctor about which ones would work best for you.

COLDS

The average adult will get about three colds a year. A cold can be caused by more than a hundred different viruses, but the symptoms are almost always the same: sore throat, runny nose, dry cough, headache, and body aches.

As miserable as a cold can make you feel, the good news is, it usually runs its course within seventy-two hours. However, a cold may quickly turn into something nasty if you ignore it. A cold is simply another way your body is telling you to slow down, at least for a few days. Cold viruses are around us every moment of every day, but we're susceptible only when our energy systems are run-down. If you were in peak condition, most of those cold viruses wouldn't have a chance against you.

If in spite of this you keep on pushing your weakened body, that cold can develop into bronchitis, pneumonia, or even worse problems. You can avoid these complications by simply stopping and taking care of your cold at the start.

Stop Running: When you have a bad cold, take the day off from work and play. Not only are you doing yourself a favor, but you're preventing the cold from spreading to those around you.

Go to Bed: Your body needs some downtime. Don't go to the gym. Don't go shopping. Don't pay your bills. Get under the covers, listen to soft music, read an uplifting book, or watch a calming TV program.

Increase Fluids: Drink a gallon of water, Gatorade, or weak tea with honey and lemon. Fenugreek tea, an herbal tea sold in health food stores, can help soothe an irritated, scratchy throat.

Steam Your Sinuses: As I noted in discussing allergies, I do not recommend decongestants or antihistamines. If your nose is stuffed, turn on a hot shower, close the bathroom door, and inhale the steam for twenty minutes at a time. It will drain your sinuses.

Let Supplements Help: There is no cure for the common cold, but you can give your immune system a boost, which will help your body defeat the infection. Take 1000 mg of vitamin C three times daily to boost immune function. Take the herb echinacea, a natural immune booster (one 500 mg capsule three times daily for a maximum of five days). Zinc lozenges, which are sold at health food stores, can also speed up your recovery.

After twenty-four hours of bed rest, you should be feeling well enough again to get up and take a walk if the weather is nice, or meet a friend for lunch. Don't overdo it! Continue to curtail your activities until the cold is gone. If you don't improve within three days, call your doctor.

DEPRESSION

Some people suffer from severe, clinical depressions that are primarily biochemical in nature. Correcting the body chemistry with state-of-the-art antidepressants like Prozac can make a huge difference in their lives.

In my experience, however, there are a great many people who are not clinically depressed, but who have a tendency to get blue every once in a while. These milder depressions are often precipitated by a life event (such as a divorce or a death in the family), physical or emotional stress, or simply a time of change. Exhaustion can also trigger mild depression. For these people, I don't believe that antidepressants are appropriate. Improved self-awareness is a far more effective and long-lasting solution.

My patient Brian is a case in point. He went off to college a reasonably happy young man, but returned six months later in a deep funk. Brian had always been somewhat moody, but until then he had usually bounced back on his own. Greatly distressing his parents, Brian said he wanted to drop out of college. They were perceptive enough to see that he was depressed, but their knee-jerk reaction was to ask for a prescription for Prozac.

Instead, I sat down with Brian to find out what was happening in his life that had pushed him into this downward spiral. Like so many college freshmen, Brian was burning the candle at both ends. Often up all night, he caught a few hours' sleep when he could in the afternoon and ate a diet that was nothing short of horrendous: Danish for breakfast washed down with a Coke. Hot dogs and french fries for lunch. Dinner was a fast-food feast. The partying (and one too many beers) on the weekend left him wrecked by Monday morning. Tired all the time, Brian felt overwhelmed by his studies. He was even too burnt-out to enjoy the social scene. No wonder he was beginning to unravel! Although he was still young and resilient, he had short-circuited his energy system by running it into the ground.

Brian's story is fairly typical of what happens to young adults when they live on their own for the first time. If their immune systems are weak, they come home from school with bronchitis or mono. Or, as in Brian's case, if their Achilles' heel is their psyche, that is precisely where they will experience the crash.

I put Brian on the Energy Pack for Sign #7 and sent him home to sleep for a few days. When I saw him the following week, his mood had improved but he seemed worried about going back to school. I didn't lecture Brian—teenagers have a way of tuning out what they don't want to hear—but I did ask him what he thought he needed to do to feel better. Interestingly, Brian was the one who made the connection between his mood and exhaustion. He also noticed how much better he felt now that he was eating normal meals at home. Brian was worried that if he returned to school, he would resume his old habits. I reassured him that that didn't have to be the case. I empowered Brian by telling them that he was now an adult and able to make conscious choices. If he chose to make his own health and well-being a top priority, he need not ever succumb to the kind of physical and mental meltdown that leads to depression.

Brian was willing to meet me halfway, but he asked for some additional help. Besides changing his sleeping and eating habits, he also wanted to try St. John's Wort, an herbal remedy for depression that is sold over the counter. I thought it was a good idea. I wasn't sure that Brian actually needed an antidepressant, but he obviously did need something to lean on. I have prescribed St. John's Wort for patients who need it, and most do well on it. Since it is a basically harmless substance, if taking St. John's Wort would allow Brian to go back to school and begin to actually enjoy college, it was fine with me.

I recently heard from Brian's parents. He's doing great—so great that he only calls home when he needs money! This story merely reinforces the certainty that you cannot separate mental symptoms from physical health—one of the major points of *Natural Energy.* It may be fashionable to prescribe a pill to cure every problem, but a quick fix is not the answer. People are stripped of the power to heal themselves, and in the process, are made to feel insignificant. The basic tools for curing the energy crisis—"Repair, Recharge, Revive"—are the same tools that were required to cure Brian's life crisis. Making a serious commitment to take care of yourself will contribute to a lifetime of good health. It will prevent the energy meltdown that is at the root of most mental and physical problems.

Life is not without stress. The key is to learn to deal with it so that the next time you encounter a similar situation, it does not derail you. While some stressful events ambush us and we are taken completely unaware, many can actually be anticipated. If you're planning a job change, living on your own for the first time, in the midst of a divorce, or even going for a job interview, understand that life changes can be depleting. Don't overload yourself with extra activities. Slow down. Be vigilant about getting enough sleep. Make sure you get enough mild exercise to relieve the tension. Eat well. Surround yourself with positive people who offer an uplifting message. Get caller ID so you don't have to talk to anyone who brings you down. Avoid loud, noisy, or upsetting environments. If you need a spiritual lift, go to church or synagogue for a weekend service, or meditate. Peace helps your energy system gather strength.

DIABETES

Diabetes, a disease characterized by an excess of sugar in the blood and urine, is a virtual epidemic in the United States. Although some people are born diabetic, most develop the disease later in life. In most cases, diabetes is entirely preventable, yet it is on the rise in the United States, the same country where exhaustion and the energy crisis are also epidemics. Why? To me, diabetes is a prime example of what happens when you run your energy system into the ground.

"Diabetes" is a general term for a group of metabolic disorders characterized by the body's inability to utilize carbohydrates, sugars, and starches found in food. There are two main types of diabetes: Type I and Type II. Type I diabetes, also known as juvenile diabetes, strikes during childhood and is caused by the failure of the pancreas to produce enough insulin. As you know, food must be broken down into smaller components before it can be used by cells. Type I diabetes is treated with diet and insulin to control blood sugar and organ damage.

Type II diabetes, also known as adult-onset diabetes, strikes later in life. Unlike Type I diabetes, where insulin is scarce, in Type II diabetes it is abundant. However, the cells have become insulin-resistant, which means that the cells don't respond to insulin.

Over time, high levels of blood sugar can lead to serious problems, including heart disease, stroke, kidney disease, neuropathy, and blindness. More than twenty-five percent of American adults develop some form of insulin resistance. Yet it is rare in the Mediterranean countries, South America, and Asia. Why?

The typical American diet must bear the blame. From our earliest days, we bombard our bodies with highly processed, oversugared, low-fiber junk. The constant barrage of sugar is more than our cells can take. They are constantly in overdrive, constantly exposed to high levels of insulin. As a result, it becomes increasingly difficult for them to break down sugar into foodstuffs that can be used to make energy. The first step toward diabetes is being overweight. It greatly increases the odds of developing insulin resistance. Don't ignore weight creep—those few extra pounds you gain each year from middle age on can add up to big problems down the road.

The sedentary lifestyle of many Americans is a major culprit. Study after study documents that we do not get enough exercise. When you don't move, you don't burn calories and you will gain weight. This slow but steady weight creep can add up to obesity very quickly, which increases the odds not only of diabetes but of cancer, heart disease, and numerous other ailments.

Adding more fiber to your diet, as I recommended in Chapter 7, will help to prevent diabetes. In fact, according to the Nurses' Health Study conducted at Harvard University, fiber may be the single most important foodstuff in protecting against insulin resistance. In the study, more than 65,000 nurses were tracked for six years. After analyzing food diaries, the researchers concluded that the women who ate the least amount of fiber were two and a half times more likely to develop insulin resistance. Eating a high-starch diet and drinking soda also greatly increased the risk of becoming insulin-resistant. If you follow the mitochondria-friendly diet, which is high in fiber and low in starchy, sugary junk food, you will greatly reduce your risk of becoming diabetic.

If adult-onset diabetes runs in your family, or if you appear to be developing high blood sugar, there are some supplements that can help:

Follow the "Repair" program for Sign #5.
Add 1000 mg of omega-3 fatty acid supplements to your diet daily.

Take 200 mcg of chromium picolinate daily. Recently, USDA researchers have found that chromium, a mineral found in food and available in supplements, can help correct insulin resistance.

Vitamin E, an antioxidant, helps cell membranes better utilize insulin. Take 400 IU of natural vitamin E daily.

DIGESTIVE TROUBLES

Food gets processed and digested in the gastrointestinal system. When the system is overwhelmed by what you put into it, it rebels; it gets backed up or it tries to rid itself of the toxins.

A breakdown in the GI system has a direct effect on the energy system. Your digestive tract is the factory that repackages food into the smaller components that can be converted into fuel by the cells. As I discussed in Chapter 7, eating the right food can enhance energy production, but eating the wrong food can clog the system. When you have difficulty digesting your food—when you have chronic gas, cramps, diarrhea, or constipation—it is a sign that you are overtaxing your GI tract and depleting your energy system.

Some people are more susceptible to stomach problems than others. For a variety of reasons, their GI systems are more vulnerable to stress and infections and may react adversely to specific foods. Usually, if people identify these problems early, they can prevent them from becoming serious. The trick is to learn to listen to your body, and follow its cues.

Nausea: The cure for occasional nausea is simple: If you feel like vomiting, don't hold yourself back. It is your body's way of trying to get rid of something it can't process. Usually, the nausea will dissipate after you vomit. Do not take any medication to stop the nausea until you are sure that your body has been given a chance to eliminate the toxins.

Diarrhea: Diarrhea is another way your body attempts to rid itself of something toxic. Don't try to stop it, but be sure that you don't become dehydrated. Drink plenty of fluids. Good choices are a sports drink (containing potassium, sugar, and salt) or miso soup. Avoid tea, which can irritate the stomach and further sap you of fluid and minerals. Eat bananas, and make sure to drink 8 oz. of fluid for each bowel movement. Take an additional 400 mg of magnesium.

Constipation: If you are constipated, drink as much as twelve glasses of water a day. It will usually do the trick. Also, increase your intake of high-fiber foods such as whole grains, fruits, and vegetables. Steer clear of heavily processed foods that are more difficult to digest. Stay away from over-the-counter laxatives! They can irritate your bowels and may cause "lazy bowel syndrome," which means you will grow dependent on them.

If you have severe symptoms or symptoms that last for more than twenty-four hours, check with your doctor. Of course, if you have a high fever, feel dizzy, are in pain, or suffer from any other unusual symptoms, don't wait. Call your doctor immediately.

If you have a tendency to develop digestive problems, watch your diet. Don't stress your system unnecessarily. Avoid huge meals, which are difficult to digest under the best of circumstances. Several smaller meals daily are better tolerated. At the first sign of stomach trouble, cut out the coffee, chocolate, and alcohol. Do not eat hot, spicy foods that can add to your misery. Most important, get enough rest. Slow down for a few days, and in all likelihood, the symptoms will disappear. Remember, if you don't have a well-functioning digestive system, you cannot get the nutrients you need to maintain your energy system. Like a house of cards, everything comes tumbling down!

✷ DIZZINESS

About one out of ten visits to the doctor is due to dizziness, a symptom which is often accompanied by nausea and/or ringing in the ears. Dizziness is a very upsetting symptom for patients because it is the embodiment of loss of control. It can strike at any time, and when it does, it can render you helpless.

Dizziness may be caused by a virus and in this case will disappear once the virus runs its course. It can also be due to an inner ear infection, sinusitis, low blood sugar, or even anxiety. Some detective work on your part may be necessary to determine the cause. Do you get dizzy when you go for long periods of time without eating? Then the likely cause is low blood sugar. Did you have a cold that may have turned into an ear infection? Is your dizziness accompanied by nausea? This could be the sign of an infection or even a bout of food poisoning.

So what to do? If you feel that your dizziness is due to either sinusitis or an ear infection, take an over-the-counter decongestant. Although I don't recommend them for colds, in this situation they can be helpful.

⚛ **Steam It Away:** Try the steam cure. If sinusitis is the problem, this will make it get better.

⚛ **Sip It Away:** Ginger capsules and tea can help reduce dizziness and quell nausea. You can buy them at your local health food store.

⚛ **Use Supplements:** Ginkgo biloba, an herb touted for its memory-enhancing properties, can help relieve dizziness by improving circulation to the brain. Ginkgo biloba capsules are sold in supermarkets, pharmacies, and health food stores. Take one 60 mg capsule three times daily. (If you are taking a blood thinner, talk to your physician before taking gingko biloba.) Many people have found that two minerals, calcium and fluoride, can relieve dizziness. No one knows why, but there is an abundance of anecdotal evidence. Combination calcium-fluoride time-release capsules are now available in health food stores and pharmacies. Take 2000 mg of vitamin C to give your immune system a boost, too, just in case you have a virus.

There are prescription drugs that may help relieve the severe nausea that may accompany dizziness. If you're suffering, call your doctor. Keep in mind that in most cases, dizziness often disappears as mysteriously as it came.

GUM DISEASE

Until recently, tooth loss was considered a normal part of growing older. Dentures were as much a rite of passage as wrinkles and gray hair. That's no longer the case. Improved nutrition, better dentistry, and fluoridated water have vastly improved the odds of keeping your teeth for a lifetime. Yet I am astonished at the number of patients who still get gum disease and require extensive periodontal work. When you understand what causes gum disease, you'll see that in most cases it is a completely preventable problem.

Gum disease is caused by the accumulation of bacteria near the gum line, which leads to chronic infection and inflammation. This inflammation can progress and destroy connective tissue and bone supporting the teeth. Tooth loss is the end result. Gum disease is a sure sign of energy system failure. Unable to fight infection and repair the damaged tissue, your body allows the teeth to die. Interestingly, in Japan, Co Q10 has been used for more than a decade as a treatment for gum disease. In fact, in one study performed at Osaka University, patients with gum disease were given either 60 mg of Co Q10 or a placebo. After eight weeks, those taking the Co Q10 showed marked improvement over the placebo takers. If you have mild gum disease, give yourself a month on the Energy Pack before submitting to further treatment. I've said it before, but I'll say it again: take care of your energy system and everything else will fall into place (and your teeth won't fall out!).

Here are some other important ways to keep your gums healthy for a lifetime.

Don't Smoke: Smoking greatly increases the risk of gum disease, not to mention cancers of the mouth.

Practice Good Hygiene: People who brush their teeth after meals and floss twice daily have healthy gums.

Avoid Sugar: A high-sugar diet promotes the growth of bacteria that will destroy your teeth and gums.

Drink Green Tea: When you eat out and can't brush, finish your meal with a cup of green tea. Green tea contains a compound that has antibacterial action against the bacteria responsible for tooth decay and gum loss. Natural chemicals in green tea called polyphenols also offer powerful protection against cancer.

HEADACHES

Headaches are a poignant way for our energy systems to force us to acknowledge our bodies. Who can ignore a headache? When your head throbs, you can't think straight and very often you can't even see straight. Your face contorts with pain and everybody immediately knows something is wrong.

Thank your energy system for this warning sign. *PAY ATTENTION!* Lie down in a dark room. Put a cold compress on your head. Take buffered aspirin or acetaminophen if you need to. Drink water. Give yourself the attention your body is asking for.

A word about aspirin: like other anti-inflammatories, it works by inhibiting the production of prostaglandins, hormones that are involved in inflammation. Although prostaglandins cause problems in other parts of the body, they actually protect the stomach lining. Since anti-inflammatories block the action of prostaglandins, they can irritate the stomach and even cause gastric ulcers and bleeding. Buffered aspirin offers some protection, but it's not foolproof. If you are prone to GI distress, avoid aspirin and other anti-inflammatories. If you can't tolerate over-the-counter anti-inflammatories, there are some new prescription ones that are gentler on the stomach. Ask your doctor about them.

A headache rarely requires a visit to the doctor. The exception to the rule is a blinding headache accompanied by high fever, nausea, or vomiting that gets worse over a few hours. In this case, definitely call your doctor. If not, try to figure out what is causing the headache on your own so that you can treat it and, in the future, prevent if from occurring in the first place.

You don't have to dig very far to figure out the cause of a headache. Most headaches are caused by insufficient sleep, stress, inflamed sinuses due to a cold or allergy, dehydration, or, last but not least, a hangover. If you think about it, you'll probably find that your headache fits into one of these categories.

Sleep It Off: There is only one cure for a headache due to exhaustion—sleep. Although some people with headaches often have difficulty sleeping, those who have headaches from lack of sleep generally fall asleep immediately if given a chance. Very often, they wake up refreshed and free of pain.

De-stress: There are various kinds of stress that can bring on headaches. For some people, emotional turmoil will result in a pounding headache. Others may find that traffic jams or sitting in front of a computer for hours at a time will trigger a headache. Learn to identify your triggers, and either avoid them or develop techniques to manage them. For example, if you know you get a headache after sitting at your desk for hours, don't allow it to happen. Get up every forty-five minutes to do a few stretches, walk around, or get some fresh air. Not only will you be avoiding a headache, but you will perform your job better. (For more tips on de-stressing, see Chapter 8, "Revive!")

Your Sinuses: If your headache occurs when you have a cold or during allergy season, inflamed sinuses may be the cause. You can easily diagnose this by closing the bathroom door, turning on a hot shower, and allowing the room to fill up with steam, as described under caring for colds. Breathe in the hot vapor for twenty minutes or so. If you feel better, it's a safe guess that your headache is sinus-related. Don't take decongestants or antihistamines; they will dry out your sinuses and cause further irritation. In most cases, you don't need to see a doctor. If your secretions are green, however, it's a sign of infection. Call your doctor; you may need an antibiotic.

Be Honest If It's a Hangover: When I was a young doctor, a man came into the ER complaining of a blinding headache and extreme nausea. Sounds like meningitis, right? One of the interns was so worried he was ready to do a spinal tap! The embarrassed man finally confessed to bingeing on eight shots of tequila the night before. What he really wanted was a cure for his hangover, but his silence almost landed him with an even worse headache! By the way, there *is* no cure for hangover. All you need is rest, a gallon of water (you're dehydrated), and aspirin.

Although it doesn't work for everyone, some people find mild exercise can help relieve a headache. Light stretching, especially in the neck and shoulder area, will relax your muscles. Avoid lifting weights or doing anything that will tense up your muscles and further aggravate your headache.

HEART DISEASE

I am always skeptical when I hear about someone having a heart attack "out of the blue." I know that in reality a heart attack is often just the final straw in what has probably amounted to years of continuous insult.

Nothing happens overnight. Typically, there are scores of little warnings that precede a heart attack but go unheeded. Throughout this book, I have explained that if we ignore the early signs of the energy crisis, our depleted energy system will break down, and leave us tired and sick. A heart attack is the quintessential breakdown of the energy system.

Having practiced medicine for twenty years, I can tell you that I have yet to meet a patient who at the age of twenty came to me and said, "I have a family history of heart disease. What can I do to prevent it?"

On the other hand, I have seen a slew of overweight, out-of-shape, fifty-something men in a state of panic. Their fathers died of early heart attacks—usually around their age—and now they are petrified of following in the fathers' footsteps. Where were these men ten years ago when their bodies first began sending them signals that they were headed for trouble?

Just because heart disease runs in your family does not mean that a heart attack is inevitable. By now everybody knows that it's important to monitor cholesterol levels and reduce your intake of saturated fat. Although these simple steps are important, to me what's even more important is raising self-awareness. If you know that you are genetically wired for heart disease, take active steps to avoid it.

Heart attacks don't come without warnings, even though we may not recognize them as such. The history of a patient who has had a heart attack "out of the blue" is often very revealing. Typically, he is under a great deal of stress, and/or smokes, or doesn't sleep well, or eats a steady diet of junk. Very often sudden-onset cardiac patients either have recently lost their jobs or are in the midst of a separation or a divorce. After I take such a history, my only surprise is that the heart attack didn't come sooner!

Invariably, I have found that these patients suffered from several if not *all* of the Seven Signs of the Energy Crisis for years before their heart attacks. They didn't recognize these warning signs or ignored them and, consequently, did nothing about them.

If you have a family history of heart disease, the earlier you intervene the better. If you've already had a heart attack, being vigilant about maintaining your health may prevent a second.

Listen when your body is telling you that you are headed for trouble. Monitor your body for the Seven Signs of the Energy Crisis. When you feel exhausted and grab for a candy bar, when you don't sleep well or are so exhausted that you stop exercising, *pay attention.* All of these problems can be handled if you catch them right away.

Following the *Natural Energy* program will greatly reduce your risk of having a heart attack. Both carnitine and Co Q10 can revitalize the mitochondria in the heart, and help to keep it strong. Throughout Europe and Japan the Energy Pack has been widely used in the prevention and treatment of heart disease for more than two decades. In the United States, cutting-edge cardiologists are now recommending it to their patients, with great results. Why wait for the first sign of heart disease? Use the Energy Pack to keep your heart healthy for your entire life.

Although the advice in "Recharge!" and "Revive!" applies to everyone, it is particularly important for people with a predisposition for heart disease. The more research that is done on heart disease, the more apparent it becomes that nutrition and lifestyle are key to prevention and treatment. It is a disease that need not follow us into the next millennium.

HYPOGLYCEMIA

I was playing tennis late one afternoon when it suddenly hit me. I felt a sinking sensation in the pit of my stomach, and I was so light-headed that I had to sit down on the court and put my head between my knees. It dawned on me that I had forgotten to eat lunch, and that my blood sugar level had gotten too low. I was suffering from a bout of hypoglycemia.

"Hypoglycemia" is a scary-sounding name for a very common condition—low blood sugar. Many people have some form of hypoglycemia, and in most cases it's not a problem. For example, women often have sudden drops in blood sugar prior to getting their periods. People who don't eat regular meals may also suffer bouts of hypoglycemia because they are trying to run their bodies on empty.

The symptoms of hypoglycemia can vary from person to person, but they often include dizziness, light-headedness, difficulty concentrating, and fatigue. Some people may feel shaky; others may get sweaty palms. The whole point is, your body is reacting this way because it is running on empty. The key is to stabilize your blood sugar levels by eating the right food at the right time.

In rare cases, however, hypoglycemia can be an early warning sign of diabetes, a problem with sugar metabolism. If you pay attention to these signs and make appropriate changes in your diet and lifestyle, you may not ever develop diabetes.

Eat Frequent, Smaller Meals: Eat a small meal or snack every three hours. Avoid refined carbohydrates and sugar; in other words, soda, chips, candy, and other junk food are off limits. Stick to lean protein and fruits and vegetables. I keep a handful of almonds and a piece of fruit in my bag so I can reach for a healthy snack when I need it.

Catch It Early: Long before you feel as if you're going to black out, most people experience some early warning signs that their blood sugar level is dropping. You may suddenly feel jittery or tired, or find yourself walking around in a brain fog. When you feel it coming on, quickly eat some protein. It will keep your blood sugar in check. Women take note: Children (and men!) often get irritable when their blood sugar is taking a dive.

Eat Before Exercise: As I learned the hard way, it's not a good idea to exercise on an empty stomach. About one hour before your exercise, have a light snack. A piece of fruit, some almonds or walnuts, or a cup of soy milk is fine. This will help prevent a sudden plunge in blood sugar when you are working out.

Take Magnesium: If you are not already doing so, add 400 mg of magnesium to your Energy Pack.

If your symptoms are severe, or you suffer from confusion or memory loss, do not wait to actually pass out: see your doctor.

INJURIES

As a physician who practiced trauma medicine, I know a great deal about injury. Most important, I know that few injuries are true acts of God or nature that strike without warning. The emergency room isn't filled with people who were caught in volcanoes, earthquakes, or even plane crashes. It is filled with people who fell asleep at the wheels of their cars, injured themselves while operating complicated machinery, slipped in the bathtub, or tripped over a piece of furniture. In reality, most injuries can be prevented. When it comes to injury, the real culprits are fatigue, inattention, and carelessness.

To me, injury is a sign that you have reached the second stage of energy system failure. You have ignored all the early warning signs and are now suffering the consequences. If you're lucky, the worst thing that will happen is a stubbed toe, a bad bruise, or a strained muscle. If you're not, you could end up falling asleep at the wheel of your car, breaking a limb, or even worse. Most serious skiing accidents, for example, occur at the end of the day, when even experienced skiers begin to tire.

Once again, the message is *pay attention.* When you are tired, don't push. Don't go for the last run. Stop while you're ahead (and in one piece).

Making a conscious effort not to run yourself into the ground will probably spare you from most serious injuries. If you eat well, rest when you're tired, and get enough sleep, you will not be accident-prone.

Despite our best efforts, minor injuries do occur. Although I always warm up before playing tennis by doing mild stretching to avoid straining a muscle, I sometimes suffer from the telltale signs of strain after a game. Treating simple injuries can be done most effectively at home. Obviously, if you're in a great deal of pain or are bleeding profusely, go to the nearest emergency room or call your doctor.

Injury can place a strain on your energy system and divert fuel that is needed to run the body into repairing and healing the wounded cells. For an added energy boost, take your Energy Pack with an additional 500 mg of carnitine and 60 mg of Co Q10 for at least three days following an injury.

Here are some other simple home remedies that will do the trick:

Bruises: A bruise or contusion is caused by the tearing of underlying blood vessels following a skin injury. Bruises can cause skin discoloration and look more serious than they actually are. A bruise will usually heal within a few days. Immediately, apply ice to the injured area to reduce pain and inflammation. Arnica ointment, a time-honored remedy which is sold in health food stores, can be applied directly to the injured area to promote healing. (Do not apply arnica to open skin.) Bromelin, an enzyme extracted from pineapple, can also help to reduce inflammation. Eat fresh pineapple or take bromelin tablets, which are available at health food stores. If you bruise easily, take vitamin C (1000 mg daily) with bioflavonoid complex. Bioflavonoids are natural chemicals, found in fruits, that can strengthen blood vessels.

Muscle Strains: Muscle strain is the most common of all sports injuries. It can usually be avoided by warming up properly before engaging in your sport, and not overexerting yourself. If you do strain a muscle, simple things can help relieve your discomfort. Rest is the most important healer. Don't beat up a strained muscle. Give it a few days to heal. As soon as possible, apply ice to the injured area. Elevate the strained limb. Use a support bandage during the day for the following two weeks to expedite healing. If you're in pain, take aspirin and apply arnica ointment to the affected area.

✷ JOINT PAIN

"My knee hurts," complained Donna, forty-six, who was annoyed that pain was interfering with her active lifestyle.

"It *really* hurts. It must be arthritis," she concluded, and then out of the clear blue sky asked, "Do you think I'll need knee replacement surgery when I'm older?"

Like Donna, most people think that joint pain automatically means you have arthritis. And even though nowadays joint replacement surgery is becoming commonplace, neither of Donna's assumptions is correct.

As I explained to her, although one of the symptoms of arthritis is joint pain, all joint pain isn't arthritis. Arthritis is characterized by inflammation, redness, and heat in the joints. You can't miss it—the joint *looks* swollen and irritated.

Joint pain, on the other hand, can be caused by many different factors. When you have the flu or are excessively tired, you may feel it in your joints. In most cases, however, simple wear and tear is the cause. It doesn't mean you need a new joint. You simply need to take better care of the old one.

I asked Donna if she was doing any new kind of exercise that might have aggravated her joints. She replied that she had purchased a treadmill the month before and might have been a bit too zealous in her workouts. In fact, right before her knee began to ache, she had increased her pace from fast walking to jogging. No wonder her knee was hurting! Our joints take an incredible amount of abuse every day. Every step you jog, you drop at least three to four times your weight on your knee joint.

Everyday activities can beat up our joints as well. We use the joints in our fingers for writing and driving, and the joints in our arms for carrying packages. Every time we walk, we are working the joints in our feet. As we get older, joints can wear out. Sometimes they get arthritic and sometimes they just get tired. They hurt for no apparent reason. More often than not, acute episodes of joint pain are caused by overuse, abuse, or misuse. Before jumping on a treadmill, riding a bike, or doing any physical activity, gently warm up your muscles by doing mild stretching. Then start the activity slowly, and gradually work up to your desired pace. Give yourself a full ten minutes to warm up. If you don't, you may injure a muscle or your joints.

At times, *underuse* can cause joint pain. If you have a sedentary job, get up and walk around for five full minutes every hour. Take a deep breath. Stretch. Relax your muscles and move your joints. Get the blood flowing. This is the best way to keep pain at bay.

If your joints really hurt, anti-inflammatories can help relieve the pain. Take buffered aspirin if it doesn't bother your stomach. Ibuprofen is fine for occasional use, as well.

If your joint pain is due to arthritis, try the new "arthritis cure"—a combination of two supplements, glucosamine and chondroitin. You can buy these supplements at any health food store, and most people find that they offer some relief.

It is most important to maintain normal weight. Carrying around excess baggage is very stressful to your joints.

What I find distressing is the fact that joint replacement surgery is becoming a major medical industry. We are fast becoming a society in which getting a new hip or knee in one's later decades is just another rite of passage. How unnecessary! Let's teach people that taking simple steps to protect their joints early in life can eliminate the need for costly and painful surgery down the road. If you do have joint pain, slow down and treat your body gently until the pain passes.

OSTEOPOROSIS

Osteoporosis is a degenerative bone disease characterized by the thinning and weakening of bone, which can lead to breaks or fractures. It is also a direct result of the slowdown in the energy system.

From childhood through adulthood, old bone is constantly being replaced with new bone in a process called remodeling. During remodeling, bone-destroying cells break down bone tissue and create microscopic cavities. Bone-building cells refill the cavities with new bone. The problem is, after age thirty, the bone breakers outpace the bone builders. Why? As the energy system loses steam, the body is unable to produce enough fuel to both run itself and make the necessary repairs. The effect of this slow but steady bone loss is cumulative.

Postmenopausal women are especially vulnerable to osteoporosis—one in three women over fifty will suffer bone fractures as a result of this disease. Although osteoporosis is perceived as a woman's disease, in reality one out of six older men will also develop the telltale thin and brittle bones.

Small-boned white and Asian women are more likely to get osteoporosis than others. The more bone you begin with, the more bone you can afford to lose. Whether you are male or female, if you have a parent with osteoporosis, you are at greater risk. If your doctor suspects that you have osteoporosis, he or she will recommend a bone density test. Periodically, he or she will monitor you for further bone loss. Fortunately, today there are several new prescription drugs that are effective treatments for osteoporosis, that stop bone loss and stimulate new bone growth.

Many factors are responsible for the slowdown in the growth of new bone. The decline in the production of key hormones is a major culprit, which is why estrogen replacement therapy is routinely prescribed for women with thinning bones. Recently, some physicians have started to prescribe testosterone to men and women to treat and prevent osteoporosis. Hormones are just one part of the osteoporosis story: nutrition and lifestyle are equally important in both the prevention and the treatment of this problem.

If you are at risk for osteoporosis, there are simple things you can do to keep your bones strong.

Eat more soy foods. In "Recharge!" I suggested adding foods such as tofu and soy milk to your diet. These foods are rich in plant estrogens, which can help preserve bone.

Take your Energy Pack! I'm not suggesting that the Energy Pack alone can prevent osteoporosis, but it can certainly help to maintain an environment that is not conducive to osteoporosis. For example, weight-bearing exercise such as walking, low-impact aerobics, and strength training can help stimulate new bone. But when you're in the throes of the energy crisis, exercise is one activity that often falls into the "I should, but don't" category. This is particularly true for Sign #1s. If taking the Energy Pack can prevent you from becoming sedentary, it will go a long way in helping to prevent bone loss.

In addition to the Energy Pack, take these supplements to maintain strong bones:

- **Calcium:** Although most people understand the importance of calcium in maintaining strong bones, few do anything about it. Calcium deficiency is still a common problem in the United States. It's simple to fix by taking 1000 mg of calcium daily with 400 IU of vitamin D to enhance calcium absorption, and eating one or two portions of calcium-rich food daily, such as yogurt or low-fat milk. Smokers and heavy drinkers (more than two drinks daily) are vulnerable to osteoporosis because cigarettes and alcohol sap the body of calcium, which is necessary to build bone. Steer clear of excessive caffeine, because it too can sap the body of calcium. A cup or two of coffee in the morning is fine, but don't overdo it. Cola drinks containing phosphates can also leach calcium from the body. Don't drink them!

Magnesium: Take 400 mg of magnesium daily. This mineral is essential for calcium metabolism and will also give you an energy boost.

Omega-3 Fatty Acids: Omega-3 fatty acids can increase the absorption of calcium by bone and decrease the excretion of calcium by the kidneys. Take 1000 mg of omega-3 fatty acid capsules daily. Again, do not use omega-3 fatty acids if you are taking a blood thinner, without checking with your doctor.

PREMENSTRUAL SYNDROME (PMS)

PMS is a natural reaction to the hormonal shifts that occur prior to menstruation. It is not an illness or a disease and should not be treated as such! PMS is just another example of how our energy systems need to be bolstered at particular times.

The typical symptoms of PMS are water retention, headache, fatigue, tender breasts, food cravings, and mild depression and/or irritability. For some women, PMS is barely noticeable. For others, however, it is so severe that it can be tantamount to having all Seven Signs of the Energy Crisis rolled into one.

The cure for PMS is simple: Listen to your body and give it what it needs.

If water retention is a problem, drink one or two cups of a natural diuretic tea called uva ursi. You can buy it at most health food stores. Since diuretics can deplete the body of magnesium and potassium, be sure to replenish those lost minerals. On the days that you need a diuretic, take 400 mg of magnesium twice daily and 100 to 200 mg of potassium once daily. Avoid high-salt foods, as they will force your body to retain fluid. Limit your water intake to six glasses of water a day.

As hormone levels start to drop, some women become so exhausted that they tend to overeat. Get out of the kitchen and take a nap. You will feel immeasurably better for it, and it doesn't put on any weight. Others experience the kind of fluctuating blood sugar levels that often lead to bingeing. To keep sugar swings in check, stick to lean protein, pasta, and fresh fruits and vegetables. In other words, follow the mitochondria-friendly food plan. Also, take your Energy Pack with 400 mg of L-glutamine to help stabilize blood sugar and 1000 mg of calcium to control mood swings and bingeing.

Evening primrose oil, an essential fatty acid, can relieve symptoms such as mood swings and tender breasts. It is also wonderful for dry skin and flaky scalp. Evening primrose oil capsules are sold in health food stores. Take 1000 mg daily.

Lately, it's become fashionable to prescribe antidepressants for PMS. I believe that antidepressants should be used only as a last resort and only for the most severe cases of PMS. In the overwhelming majority of cases, my simple steps will work wonders.

SKIN PROBLEMS

I can always tell when my friend Deborah is upset. I can see it on her face—literally. Practically overnight, her usual peaches-and-cream complexion is covered with angry red bumps. Deborah is forever trying to treat her breakouts with different creams and potions, but the fact is, she is wasting her time. You can't heal skin from the outside in; skin care begins from the inside out.

Although we think of it as having a purely decorative function, skin is one of the most important organs in the body. Like the heart, the liver, the spleen, and the kidneys, skin has specific jobs. It forms a physical barrier between us and the outside world. It protects us against bacteria and foreign objects, and helps to maintain body temperature. Like other organs, skin responds to its environment. If you are stressed out, eating a poor diet, and not drinking enough water, it will be reflected on your skin.

I'm not saying that external assaults don't play a role in skin health. By now, everyone knows that baking for hours in the sun not only will cause wrinkles but will promote skin cancer. Just remember that our internal environment is equally important, and what's happening on the inside will be reflected in what's happening on the outside.

When people are mired in the energy crisis, it's written all over their faces. Their complexions are not pink and glowing but sallow and dull. Typically, they have dark circles under their eyes. When the energy system is melting down, it's impossible for the body to maintain its cells properly, on the inside or the outside of the body.

Why do some teenagers get bad acne while others do not? Certainly, genetics and hormonal fluctuations are one factor in acne, but they are not the whole story. Personal hygience and eating habits are also important. All of these factors influence the health and appearance of skin.

The impact of physical and emotional stress on skin is often underestimated. For example, poison ivy can spread like wildfire in allergic people who are under a great deal of stress.

Here are some more tips on maintaining healthy, beautiful skin:

Dry Skin: Moisturizing creams can help make skin look and feel better, at least temporarily. However, when I see patients with very dry skin or a dry, flaky scalp, it is a sign to me that they need more essential fatty acids in their diet. I recommend 1000 mg of evening primrose oil daily, as well as more omega-3-rich foods, such as salmon, albacore tuna, and sardines. People with dry skin should be well hydrated, eat a healthy diet, and make sure they have normal thyroid function.

Oily Skin: People with oily skin are prone to develop clogged pores, blackheads, and pimples. Often, they overreact by using harsh cleansers to rid their faces of these blemishes. This can be a mistake. Recently, a patient of mine used one of the new blackhead treatments which involve placing an adhesive strip over the nose. When she tore the strip off, what remained was an angry red splotch! Although these products may be fine for some people, she had developed an awful allergic reaction. So be careful. Before using any product on your face, test it on a small portion of skin on your upper arm. Cover the treated area with a Band-Aid. If you don't have a reaction to the product, you can then use it on your face. Treat your skin gently. To keep oil under control, wash your face several times a day, but use gentle cleansers. My favorites are soaps that contain chamomile extract. They do the job well, but they do it without irritating the skin. A mild astringent can also help to keep oil under control. Avoid fried or greasy food. I know that dermatologists say the oil content of your diet makes no difference in terms of skin quality, but I don't believe it. When I eat an oily meal like pizza, I can literally feel the oil seeping through my skin.

Aging Skin: Limiting sun exposure is one way to prevent premature aging of your skin. You know the drill: Stay out of the sun during peak burning times (ten a.m. to three p.m.), and wear a hat and sunscreen of at least SPF 15 when you do go out. Eat lots of fresh fruits and vegetables. They contain vitamins, minerals, and phytochemicals that can help counteract the destructive effect of UV rays from the sun. I use vitamin C cream on my skin because it stimulates the formation of collagen, the underlying structure that holds skin together. Loss of collagen contributes to skin aging.

The Energy Pack will also keep your skin healthier and more youthful. By bolstering the energy system, you are giving your skin the ability to produce new cells and heal damaged cells more quickly. The positive things you are doing on the inside of your body will be reflected on the outside.

Chapter Ten

ENERGIZED FOREVER

In Chapter 1, I asked you to imagine what it would be like to greet each day full of energy and ready to take on the world. I asked you to contemplate how good it would feel to face the daily challenges of life with renewed vigor and vitality.

By now, you know that it feels wonderful.

By following the *Natural Energy* program, you have gone from tired to terrific in no time. Your energy levels are high, your mood is soaring, and you look as good as you feel.

And getting there was so very easy. Take care of your energy system and it will take care of you. Once you understand that the energy system is the most important system of all, a whole new world of possibilities opens up.

We now know that if you keep your energy system strong, everything else falls into place. The reverse is also true—neglect your energy system and everything else falls apart.

The energy crisis tells you that your energy system is slowing down. Do nothing to stop the decline, and it will affect every other system in the body. When your mitochondria begin to wear out, your cells cannot make enough fuel to run the body. Nor can they repair injured cells, remove toxins, or make new cells efficiently. Without adequate fuel, brain cells can't think, heart cells can't beat, immune cells can't fight infection, and so on. Like a stack of dominoes, one by one, each system topples.

This dismal scenario need not ever happen—not to you, not to me, not to anyone. The *Natural Energy* program revitalizes the energy system quickly and effectively, and I have seen it work time and time again. The transformation in my patients is nothing short of astounding. Their energy levels are off the charts, and once they are rescued from their personal energy crises, they begin to enjoy life again. They are no longer too tired to do the things that make their lives rich and meaningful. They are optimistic about the future, not wistful about the past.

They can look forward to living a long, vibrant life in a strong, healthy body, as can you.

For the first time we can say with confidence that illness is not an inevitable part of life, nor a consequence of growing older. It is a direct result of energy system failure. We get sick because our energy stores begin to wear down and we are no longer able to combat infection, repair wounds efficiently, replace worn-out cells, or defeat cancer cells before they can spread.

The old approach has been to tackle this problem one ailment at a time *after the fact.* Wait until the heart gives out, and when it does, patch it up with bypass surgery. When a hipbone breaks, replace it with a new hip joint. When diabetes sets in, prescribe drugs to control blood sugar levels. We keep trying to prevent and cure *disease,* not promote and maintain *health.* Until now, that is. Support the master system running the body, the energy system; keep it healthy and vibrant and strong, and you will stop those other problems from occurring in the first place.

Breathtakingly simple, isn't it?

I'm not suggesting that all illness is a thing of the past. There are other factors, like genetics and environment, that are not within our control. Inheriting the wrong gene, exposure to toxins in the air and water, or extreme stress can all increase the odds of getting sick down the road. But even when the deck is stacked against us, we can land an ace. We as individuals have a great deal more power to control our destinies than we ever imagined.

Natural Energy teaches you how to tap into that power, the power to be the keeper of your health and your energy system. The three simple rules of *Natural Energy*—"Repair, Recharge, Revive"—give you the tools to live life to its fullest, from your earliest years through your last decades. "Repair" shows how the Energy Pack can keep your mitochondria youthful. "Recharge" teaches the difference between bad food that clogs the energy system and good food that produces high-quality fuel. "Revive" alerts you to the everyday energy zappers that can sap you of your natural energy.

The *Natural Energy* program fills an information gap between doctors and patients that is growing by leaps and bounds. For years, the practice of medicine has been dominated by doctors and disease. Although we have paid lip service to preventive medicine, the fact is, medicine has revolved around the treatment of seemingly unrelated ailments. We spring into action when a patient has a tangible, diagnosable disease, but we are lethargic about helping patients maintain health.

Let me make one point very clear. I am not doctor-bashing—a practice which has become fashionable these days—

nor am I disregarding the enormous amount of good that has been accomplished by modern medicine. No one who has worked in trauma medicine as I have can dismiss the miracles that occur every day in the ER. When it comes to repairing a mangled body or resuscitating a patient who has had a heart attack, we are the best in the world.

But outside the hospital, the balance of power has shifted dramatically. There's one fact of life that's difficult for doctors to accept: We doctors need our patients as much as they need us. I have learned more by listening to my patients than I did throughout my four years of medical school. Patients who are very attuned to their bodies can breathe life into the dull descriptions of symptoms listed in a medical text. Not that patients are always right either. You can also learn from patients who are out of touch with their bodies.

Natural Energy grants you a power you have always had—to be the keeper of your own health. Long before you are officially sick, your body signals you that something is wrong. Before you feel the first chest pain of heart disease, find yourself battling cancer, or need a joint replacement, your body is crying out for help.

Sometimes the signals are ridiculously simple to understand. Those of you who smoke can remember that when you inhaled your first cigarette, you started to choke. Those who listened to that cough, and didn't light the next cigarette, avoided early heart attacks and lung cancer. Many of those who ignored the body's clear message were not as lucky.

The Seven Signs of the Energy Crisis teach you how to decode your body's early warning signals in time to make positive changes. Each sign gives you a chance to rewrite your personal history; gives you the power to set things right.

When Sign #1 strikes, for example—when your brain is telling you to exercise, but your body is saying to forget it—it's time to reevaluate your lifestyle. Ask yourself why you feel the way you feel. Are you sleeping enough? Are you eating correctly? Or are you running yourself ragged doing so many unnecessary things that you don't have energy left to do what's truly important? If you change your destructive course early enough, if you repair, recharge, and revive, your condition will not deteriorate to the point at which you are a walking risk factor for virtually every disease from obesity to cancer to diabetes.

Natural Energy marks the beginning of a revolution that will forever change how we view our bodies. It provides a new paradigm for the way medicine will be practiced in the twenty-first century. The potential in terms of enhancing human health and well-being is nothing short of spectacular.

Natural Energy changes our basic assumptions about what it means to be truly healthy. Long before we had fancy, high-tech diagnostic tests, practitioners of traditional medicine relied on how patients looked and felt to determine their overall health. If a patient complained of fatigue or looked tired, they would assume something was wrong. Today, diagnosis has become a high-tech game utterly removed from common sense. If your lab tests say you are healthy, you are healthy regardless of how you feel. In fact, if a patient's lab tests don't conform with his complaints, most doctors believe the lab tests and dismiss the patient's problem as "all in his head." At the same time, the impact of lifestyle has been minimized. Doctors often dismiss questions about nutrition with a pat, "Eat a well-balanced diet," without ever saying what that is. And patients infer—correctly—that the doctor doesn't give a damn about nutrition, so why should they? Supplements? It's a small though growing minority of doctors who even know the basic research about them. In this arena, patients are often better informed than their doctors!

Natural Energy reinforces the traditional belief that how we feel is more relevant to our health than a desk piled high with lab results. It puts people back in charge of their own health care—and ultimately their own destinies. It puts doctors back into their role as healers, but this time in partnership with their patients.

Natural Energy is the first book to be written about the energy system, but I doubt it will be the last. Scientists have only begun to unravel the mystery of the energy system, and almost daily, new studies are being published that reinforce the powerful role it plays in the body. I look forward to some spectacular advances in mitochondrial medicine in the next century. In years to come, I expect to see the power of the energy system tapped to produce new therapies for hard-to-treat ailments such as Alzheimer's, cancer, and heart failure.

Innovative physicians have already demonstrated that boosting the energy system can be a powerful therapeutic tool. For example, Peter Langsjoen, the cardiologist in Tyler, Texas, has used Co Q10 to save the lives of people with incurable heart failure, some given only days to live. The late Dr. Karl Folkers, a pioneer in Co Q10 research, found it was an effective treatment for advanced breast cancer (along with other conventional therapies). Several landmark studies have already shown that carnitine supplements can slow down the progression of Alzheimer's disease. Much as antibiotics revolutionized the treatment of disease in the twentieth century, these new energy boosters offer hope about diseases of aging that until now had appeared to be intractable.

Bone fractures are another casualty of mitochondrial burnout that will one day be treated more effectively by bolstering the energy system. Throughout this book, I have used the example of a hip fracture to show the full impact of the energy crisis. As you know, when a young person suffers a bone injury, it usually heals rapidly. But the same injury can cripple or even kill an older person. Why? Because an old energy system cannot heal a bone fracture as readily as a youthful energy system can. The same principle holds true for osteoporosis, a potentially debilitating disease that affects one out of four postmenopausal women. In osteoporosis, old bone is worn down more rapidly than it is replaced by new bone. Osteoporosis is a classic example of energy system default. It doesn't happen in young people because their energy systems are still strong. I believe maintaining a vigorous energy system for your entire life will have a strong protective effect against these diseases, but finding a way to revitalize a severely weakened energy system may be the silver bullet to cure these and countless other problems.

Since I often advocate natural remedies like the Energy Pack, patients ask me if I am an alternative doctor. I shun these labels because they are meant to box doctors into one kind of medical practice, and that only works to the disservice of my patients. I use whatever works best for a patient, whether it's a prescription drug, a supplement, or a nutritional approach. The only kind of medicine I admit to practicing is commonsense medicine.

As I approach my fifth decade with enthusiasm and optimism, I have equally high hopes for my patients. I come by my optimism naturally. I am part of a unique generation of men and women that has changed society every step of the way. We baby boomers have flatly refused to quietly accept things as they are; rather, we are leaving our mark on every institution in society. For the most part, these changes have been overwhelmingly positive. If baby boomers had not changed traditional sex role stereotypes, I would probably not have been admitted to medical school or be practicing today in a field that I love. The fact that baby boomers refuse to accept aging lying down has shattered basic assumptions about what it means to grow old. Most of us expect to be here well into our late decades, not just surviving but living full and interesting lives.

Natural Energy is insurance against the downward spiral of disease and debility that results when you neglect your energy system. *Natural Energy* provides an easy-to-follow blueprint for reinvigorating your body and your mind, not just for the short term but for years to come. *Natural Energy* is not just about feeling good for the moment, but about fortifying your body so that you feel good today, tomorrow, and ten years from now.

Use it in good health.

ERIKA SCHWARTZ, M.D.
IRVINGTON, NEW YORK
JANUARY 1999

Selected Bibliography

Ahmad, S., Robertson, H. T., Golper, T. A., et al. "Multicenter Trial of L-Carnitine in Maintenance Hemodialysis Patients. II. Clinical and Biochemical Effects." *Kidney International* 38:912–18. 1990.

Ames, B. N., Shigenaga, M. K., and Hagen, T. M. "Oxidants, Antioxidants, and the Degenerative Diseases of Aging." *Proceedings of the National Academy of Sciences of the United States of America* 90:7915–22. 1993.

Amimoto, T., Matsura, T., Koyama, S., et al. "Acetaminophen-Induced Hepatic Injury in Mice: The Role of Lipid Peroxidation and Effects of Pretreatment with Coenzyme Q10 and α-Tocopherol." *Free Radical Biology and Medicine* 19(2):169–76. 1995.

Arenas, J., Huertas, R., Campos, Y., et al. "Effects of L-Carnitine on the Pyruvate Dehydrogenase Complex and Carnitine Palmitoyl Transferase Activities in Muscle of Endurance Athletes." *FEBS Letters* 341:91–3. 1994.

Arsenian, M. A. "Carnitine and Its Derivatives in Cardiovascular Disease." *Progress in Cardiovascular Diseases* 40(3):265–86. 1997.

Atar, D., Spiess, M., Mandinova, A., et al. "Carnitine—From Cellular Mechanisms to Potential Clinical Applications in Heart Disease." *European Journal of Clinical Investigation* 27:973–6. 1997.

Baggio, E., Gandini, R., Plancher, A. C., et al. "Italian Multicenter Study on the Safety and Efficacy of Coenzyme Q10 as Adjunctive Therapy in Heart Failure." *Molecular Aspects of Medicine* 15:s287:s294. 1994.

Bandy, B., and Davidson, A. J. "Mitochondrial Mutations May Increase Oxidative Stress: Implications for Carcinogenesis and Aging?" *Free Radical Biology and Medicine* 8:523–39. 1990.

Bartels, G. L., Remme, W. J., and Scholte, H. R. "Acute Myocardial Ischaemia Induces Cardiac Carnitine Release in Man." *European Heart Journal* 18:84–90. 1997.

Beal, F., and Littarru, G. P. "Coenzyme Q10 as a Potential Treatment for Neurodegenerative Diseases." *The First Conference of the International Coenzyme Q10 Association Programme and Abstracts. Boston, MA,* May 21–24, 1998.

Beal, M. F., Henshaw, D. R., Jenkins, B. G., et al. "Coenzyme Q10 and Nicotinamide Block Striatal Lesions Produced by the Mitochondrial Toxin Malonate." *Annals of Neurology* 36(6):882–8. 1994.

Bertelli, A., Bertelli, A. A. E., Giovannini, L., et al. "Protective Synergic Effect of Coenzyme Q10 and Carnitine on Hyperbaric Oxygen Toxicity." *International Journal of Tissue Reactions* 12(3):193–6. 1990.

Bertelli, A., Cerrati, A., Giovannini, L., et al. "Protective Action of L-Carnitine and Coenzyme Q10 Against Hepatic Triglyceride Infiltration Induced by Hyperbaric Oxygen and Ethanol." *Drugs Under Experimental and Clinical Research* 19(2):65–8. 1993.

Bertelli, A., Ronca, F., Ronca, G., et al. "L-Carnitine and Coenzyme Q10 Protective Action Against Ischaemia and Reperfusion of Working Rat

Heart." *Drugs Under Experimental and Clinical Research* 18(10):431–6. 1992.

Block, L. H., Georgopoulos, A., Mayer, et al. "Nonspecific Resistance to Bacterial Infections; Enhancement by Ubiquinone-8." *Journal of Experimental Medicine* 148:1228–40. 1978.

Bohles, H., Noppeney, T., Akcetin, Z., et al. "The Effect of Preoperative L-Carnitine Supplementation on Myocardial Metabolism During Aorto-Coronary-Bypass Surgery." *Current Therapeutic Research* 39(3):429–35. 1996.

Brass, E. P., and Hiatt, W. R. "Minireview: Carnitine Metabolism During Exercise." *Life Sciences* 54(19):1383–93. 1994.

Breningstall, G. N. "Carnitine Deficiency Syndromes." *Pediatric Neurology* 6(2):75–81. 1990.

Cacciatore, L., Cerio, R., Ciarimboli, M., et al. "The Therapeutic Effect of L-Carnitine in Patients with Exercise-Induced Stable Angina: A Controlled Study." *Drugs Under Experimental and Clinical Research* 17(4):225–35. 1991.

Capurso, A., Resta, F., Colacicco, M., et al. "Effect of L-Carnitine on Elevated Lipoprotein(a) Levels." *Current Therapeutic Research* 56(12): 1247–53. 1995.

Chopra, R., Goldman, R., and Bhagavan, H. N. "A New Coenzyme Q10 Preparation with Enhanced Bioavailability." *FASEB Journal* 11:A586. 1997.

Combs, A. B., Choe, J. Y., Truong, D. H., et al. "Reduction by Coenzyme Q10 of the Acute Toxicity of Adriamycin in Mice." *Research Communications in Chemical Pathology and Pharmacology* 18(3):565–8. 1977.

Coulter, D. L. "Carnitine Deficiency in Epilepsy: Risk Factors and Treatment." *Departments of Pediatrics and Neurology, Boston University School of Medicine, and the Division of Pediatric Neurology, Boston City Hospital, Boston, MA.* 1995.

Demeyere, R., Lormans, P., Weidler, B., et al. "Cardioprotective Effects of Carnitine in Extensive Aortocoronary Bypass Grafting: A Double-Blind, Randomized, Placebo-Controlled Clinical Trial." *International Anesthesia Research Society* 71:520–8. 1990.

DeSimone, C., and Martelli, E. A. "Stress, Immunity and Ageing: A Role for Acetyl-L-Carnitine." Elsevier Science Publisher B.V. New York, NY. 1989.

Digiesi, V., Cantini, F., Oradei, A., et al. "Coenzyme Q10 in Essential Hypertension." *Molecular Aspects of Medicine* 15:s257–63. 1994.

Ferrari, R., DiMauro, S., and Sherwood, G. "L-Carnitine and Its Role in Medicine: From Function to Therapy." Academic Press. San Diego, CA. 1992.

Folkers, K., Hanioka, T., Xia, L. J., et al. "Coenzyme Q10 Increases T4/T8 Ratios of Lymphocytes in Ordinary Subjects and Relevance to Patients Having the AIDS Related Complex." *Biochemical and Biophysical Research Communications* 176(2):786–91. 1991.

Folkers, K., Langsjoen, P., Willis, R., et al. "Lovastatin Decreases Coenzyme Q Levels in Humans." *Proceedings of the National Academy of Sciences of the United States of America* 87(22):8931–4. 1990.

Folkers, K., Manabu, M., and McRee, J. Jr. "The Activities of Coenzyme Q10 and Vitamin B4 for Immune Responses." *Biochemical and Biophysical Research Communications* 193(1):88–92. 1993.

Folkers, K., Osterborg, A., Nylander, M., et al. "Activities of Vitamin Q10 in Animal Models and a Serious Deficiency in Patients with Cancer." *Biochemical and Biophysical Research Communications* 234(2):296–9. 1997.

Folkers, K., Vadhanavikit, S., and Mortenses, S. A. "Biochemical Rationale and Myocardial Tissue Data on the Effective Therapy of Cardiomyopathy with Coenzyme Q10." *Proceedings of the National Academy of Sciences of the United States of America* 82:901–4. 1985.

Franceschi, C., Monti, D., Troiano, L., et al. "Down Syndrome, Centenarians and Oxidative Stress: The Rationale for the Use of Nioctinamide and Carnitines as Anti-Aging Substances." *Bulletin of Molecular and Biological Medicine* 18:145–67. 1993.

Gabauer, I., Pechan, I., Fischer, et al. "Cardioprotection by L-Carnitine in Patients During Coronary Surgery." *Journal of Molecular and Cellular Cardiology* 29(5):s12. 1997.

Gaby, A. R. "Coenzyme Q10." Pizzorno, Murray & Gaby. 1992.

Ghirardi, O., Peschechera, A., Ramacci, M. T., et al. "Long-Term Administration of L-Carnitine Modifies the Coenzyme Q Pattern in the Myocardium and Skeletal Muscle of the Old Rat." *International Society for Myochemistry Second General Congress.* 1997.

Giovnali, P., Fenocchio, D., Montanari, G., et al. "Selective Trophic Effect of L-Carnitine in Type I and IIa Skeletal Muscle Fibers." *Kidney International* 46:1616–19. 1994.

Guarnieri, G., Panzetta, G., and Toigo, G. "Amelioration of Cardiac Function by L-Carnitine Administration in Patients on Haemodialysis." *Contributions to Nephrology* 98:28–35. 1992.

Hagen, T. M., Ingersoll, R. T., Wehr, Carol M., et al. "Acetyl-L-Carnitine Fed to Old Rats Partially Restores Mitochondrial Function and Ambulatory Activity." *Proceedings of the National Academy of Sciences of the United States of America* 95:9562–6. 1998.

Hagen, T. M., Yowe, D. L., Bartholomew, J. C., et al. "Mitochondrial Decay in Hepatocytes From Old Rats: Membrane Potential Declines, Heterogeneity and Oxidants Increase." *Proceedings of the National Academy of Sciences of the United States of America* pp. 3–15. 1997.

Hanioka, T., Tanaka, M., Ojima, M., et al. "Effect of Topical Application of Coenzyme Q10 on Adult Periodontitis." *Molecular Aspects of Medicine* 15:s241–8. 1994.

Harman, D. "The Biologic Clock: The Mitochondria?" *Journal of the American Geriatrics Society* 20(4):145–7. 1972.

Ido, Y., McHowat, J., Chang, K. C., et al. "Neural Dysfunction and Metabolic Imbalances in Diabetic Rats." *Diabetes* 43:1469–77. 1994.

Iliceto, S., Scrutinio, D., Bruzzi, P., et al. "Effects of L-Carnitine Administration on Left Ventricular Remodeling After Acute Anterior Myocardial Infarction: The L-Carnitine Ecocardiografia Digitalizzata Infarto Miocardico (CEDIM) Trial." *Journal of the American College of Cardiology* 26(2):380–7. 1995.

Jacota, K. G. C., Abarquez, Jr., G. O., Topacio, M. T., et al. "Effect of L-Carnitine on the Limitation of Infarct Size in One-Month Postmyocardial Infarction Cases: A Multicentre, Randomised, Parallel, Placebo-Controlled Trial. *Clinical Drug Investigation* 11(2):1996.

Kamikawa, T., Kobayashi, A., Yamashita, T., et al. "Effects of Coenzyme Q10 on Exercise Tolerance in Chronic Stable Angina Pectoris." *The American Journal of Cardiology* 56:247–51. 1985.

Kishi, T., Kishi, H., Watanabe, T., et al. "Bioenergetics in Clinical Medicine. XI. Studies on Coenzyme Q and Diabetes Mellitus." *Journal of Medicine* 7(3,4):307–21. 1976.

Kuklinski, B., Weissenbacher, E., and Fahnrich, A. "Coenzyme Q10 and Antioxidants in Acute Myocardial Infarction." *Molecular Aspects of Medicine* 15:s143–7. 1994.

Langsjoen, H., Langsjoen, P., Langsjoen, P., et al. "Usefulness of Coenzyme Q10 in Clinical Cardiology: A Long-term Study." *Molecular Aspects of Medicine* 15:s165–75. 1994.

Langsjoen, P., Langsjoen, P., Willis, R., et al. "Treatment of Essential Hypertension with Coenzyme Q10." *Molecular Aspects of Medicine* 15: s264–72. 1994.

Langsjoen, P., Vadhanavikit, S., and Folkers, K. "Response of Patients in Classes III and IV of Cardiomyopathy to Therapy in a Blind and Crossover

Trial with Coenzyme Q10." *Proceedings of the National Academy of Sciences of the United States of America* 82:4240–4. 1985.

Langsjoen, P. H., Folkers, K., Lyson, K., et al. "Pronounced Increase of Survival of Patients with Cardiomyopathy When Treated with Coenzyme Q10 and Conventional Therapy." *International Journal of Tissue Reactions* 12(3):163–8. 1990.

LaPlante, A., Vincent, G., Poirier, M., et al. "Effects and Metabolism of Fumarate in the Perfused Rat Heart. A 13C Mass Isotopomer Study." *The American Physiological Society* pp.E74–E82. 1997.

Laschi, R., "L-Carnitine and Ischaemia; A Morphological Atlas of the Heart and Muscle." *Biblioteca Scientifica.* 1987.

Lockwood, K., Moesgaard, S., and Folkers, K. "Partial and Complete Regression of Breast Cancer in Patients in Relation to Dosage of Coenzyme Q10." *Biochemical and Biophysical Research Communications* 199(3):1504–8. 1994.

Lockwood, K., Moesgaard, S., Hanioka, T., et al. "Apparent Partial Remission of Breast Cancer in 'High Risk' Patients Supplemented with Nutritional Antioxidants, Essential Fatty Acids and Coenzyme Q10." *Molecular Aspects of Medicine* 15:s231–40. 1994.

Lockwood, K., Moesgaard, S., Yamamoto, T., et al. "Progress on Therapy of Breast Cancer with Vitamin Q10 and the Regression of Metastases." *Biochemical and Biophysical Research Communications* 212(1):172–7. 1995.

Luppa, D., and Loster, H. "L-Carnitine Through Urine and Sweat in Athletes in Dependence on Energy Expenditure During Training." *Ponte Press Bochum* pp. 278–9. 1996.

Marconi, C., Sassi, G., Carpinelli, A., et al. "Effects of L-Carnitine Loading on the Aerobic and Anaerobic Performance of Endurance Athletes." *European Journal of Applied Physiology* 54:131–5. 1985.

Mariani, J., Ou, R., Nagley, P., et al. "Impaired Recovery of Aged Human Myocardium After Ischemia and Hypoxia Correlates with Increased Mitochondrial DNA Deletions." *Cardiac Surgical Research Unit, Baker Medical Research Institute and Alfred Hospital; Department of Biochemistry and Molecular Biology, Monash University; Centre for Molecular Biology and Medicine, Epworth Hospital, Melbourne, Australia.*

Mecocci, P., MacGarvey, U., Kaufman, A. E., et al. "Oxidative Damage to Mitochondrial DNA Shows Marked Age-Dependent Increase in Human Brain." *Annals of Neurology* 34(4):609–16. 1993.

Moretti, S., Alesse, E., Di Marzio, L., et al. "Effect of L-Carnitine on Human Immunodeficiency Virus-1 Infection-Associated Apoptosis: A Pilot Study." *Blood* 91(10):3817–24. 1998.

Nakamura, R., Littarru, G. P., Folkers, K., et al. "Study of Co Q10-Enzymes in Gingiva from Patients with Periodontal Disease and Evidence for a Deficiency of Coenzyme Q10." *Proceedings of the National Academy of Sciences of the United States of America* 71:1456–60. 1974.

Ogura, R., Toyama, H., Shimada, T., et al. "The Role of Ubiquinone (Coenzyme Q10) in Preventing Adriamycin-Induced Mitochondrial Disorders in Rat Heart." *Journal of Applied Biochemistry* 1:325–35. 1979.

Paradies, G., Ruggiero, F. M., Petrosillo, G., et al. "Carnitine-Acylcarnitine Translocase Activity in Cardiac Mitochondria From Aged Rats: The Effect of Acetyl-L-Carnitine." *Mechanisms of Ageing and Development* 84:103–12. 1995.

Paradies, G., Ruggiero, F. M., Petrosillo, G., et al. "Effect of Aging and Acetyl-L-Carnitine on the Activity of Cytochrome Oxidase and Adenine Nucleotide Translocase in Rat Heart Mitochondria." *FEBS Letters* 350:213–15. 1994.

Parnetti, L., Abata, G., Bartorelli, L., et al. "Multicentre Study of l-α-Glyceryl-Phosphorylcholine *vs* ST200 among Patients with Probable Senile Dementia of Alzheimer's Type." *Drugs & Aging* 3(2):159–64. 1993.

Paulson, D. J. "Carnitine Deficiency-Induced Cardiomyopathy." *Molecular and Cellular Biochemistry* 180:33–41. 1998.

Pearl, J. M., Hiramoto, J., Laks, H., et al. "Fumarate-Enriched Blood Cardioplegia Results in Complete Functional Recovery of Immature Myocardium." *The Society of Thoracic Surgeons* 57:1636–41. 1993.

Pettegrew, J. W., Klunk, W. E., Panchalingam, K., et al. "Clinical and Neurochemical Effects of Acetyl-L-Carnitine in Alzheimer's Disease." *Neurobiology of Aging* 16(1):1–4. 1995.

Plioplys, A., and Plioplys, S. "Amantadine and L-Carnitine Treatment of Chronic Fatigue Syndrome." *Neuropsychobiology* 35:16–23. 1997.

Pola, P., Tondi, P., Dal Lago, A., et al. "Statistical Evaluation of Long-Term L-Carnitine Therapy in Hyperlipoproteinaemias." *Drugs Under Experimental and Clinical Research* 9(12):925–34. 1983.

Rebouche, C. J. "Carnitine Metabolism and Function in Humans." *Annual Review of Nutrition* 6:41–66. 1986.

Ronca-Testoni, S., Zucchi, R., Ronca, F., et al. "Effect of Carnitine and Coenzyme Q10 on the Calcium Uptake in Heart Sarcoplasmic Reticulum of Rats Treated with Anthracyclines." *Drugs Under Experimental and Clinical Research* 18(10):437–42. 1992.

Rosenfeldt, F. L., Pepe, S., Ou, R., et al. "Coenzyme Q10 Improves the Tolerance of the Senescent Myocardium to Aerobic and Ischemic Stress: Studies in Rats and in Human Atrial Tissue." *International Coenzyme Q10 Association, Boston, MA.* 1998.

Rossi, C. S., and Siliprandi, N. "Effect of Carnitine on Serum HDL-Cholesterol: Report of Two Cases." *The Johns Hopkins Medical Journal* 150:51–4. 1982.

Rowland, M. A., Nagley, P., Linnane, A. W., et al. "Coenzyme Q10 Treatment Improves the Tolerance of the Senescent Myocardium to Pacing Stress in the Rat." *Cardiac Surgical Research Unit, Baker Medical Research Institute, Prahran, Victoria, Australia; Department of Biochemistry and Molecular Biology, Monash University, Clayton, Victoria, Australia; Centre for Molecular Biology and Medicine, Richmond, Victoria, Australia.*

Schulz, J. B., Henshaw, D. R., Matthews, R. T. "Coenzyme Q10 and Nicotinamide and a Free Radical Spin Trap Protect Against MPTP Neurotoxicity." *Experimental Neurology* 132:279–83. 1995.

Seim, H., and Loster, H. "Carnitine: Pathobiochemical Basics and Clinical Applications." *Ponte Press Bochum*. 1996.

Shannon, D. W., and Wolfe, G. S. "Carnitine Deficiency: A Missed Diagnosis." *The Journal of Care Management* 3(3):12–24. 1997.

Shigenaga, M. K., Hagen, T. M., and Ames, B. N. "Oxidative Damage and Mitochondrial Decay in Aging." *Proceedings of the National Academeny of Sciences* 91:10771–8. 1994.

Simonetti, S., Chen, X., DiMauro, S., et al. "Accumulation of Deletions in Human Mitochondrial DNA During Normal Aging: Analysis by Quantitative PCR." *Biochemica et Biophysica Acta* 1180:113–22. 1992.

Sinatra, S. T. "Coenzyme Q10: A Vital Therapeutic Nutrient for the Heart with Special Application in Congestive Heart Failure." *Connecticut Medicine* November, 1997.

Singh, R. B., Niaz, M. A., Agarwal, P., et al. "A Randomised, Double-Blind, Placebo-Controlled Trial of L-Carnitine in Suspected Acute Myocardial Infarction." *Postgraduate Medical Journal* 72:45–50. 1996.

Sugiyama, S., Yamada, K., and Ozawa, T. "Preservation of Mitochondrial Respiratory Function by Coenzyme Q10 In Aged Rat Skeletal Muscle." *Biochemistry and Molecular Biology International* 37(6):1111–20. 1995.

Suzuki, M., Kanaya, M., Mukamatsu, S., et al. "Effects of Carnitine Administration, Fasting, and Exercise on Urinary Carnitine Excretion in Man." *Journal of Nutritional Science and Vitaminology* 22:169–74. 1976.

Thein, L. A., Thein, J. M., and Landry, G. L. "Ergogenic Aids." *Physical Therapy* 75(5):426–39. 1995.

Tritschler, H. J., Andreeta, F., Moraes, C. T., et al. "Mitochondrial Myopathy of Childhood Associated with Depletion of Mitochondrial DNA." *Neurology* 42:209–17. 1992.

Uhlenbruck, G. "L-Carnitine and the Immune System: From the Mode of Metabolism to the Modulation of Membranes." *Ponte Press Bochum* pp. 47–60. 1996.

Virmani, M. A., Biselli, R., Spadoni, A., et al. "Protective Actions of L-Carnitine and Acetyl-L-Carnitine on the Neurotoxicity Evoked By Mitochondrial Uncoupling or Inhibitors." *Pharmacological Research* 32(6):383–9. 1995.

Wallace, D. C., "Mitochondrial Genetics: A Paradigm for Aging and Degenerative Diseases?" *Science* 256:628–32. 1992.

Watanabe, S., Ajisake, R., Masuoka, T., et al. "Effects of L-and DL-Carnitine on Patients with Impaired Exercise Tolerance." *Japanese Heart Journal* 36(3):319–31. 1995.

Wyss, V., Ganzit, G. P., and Rienzi, A. "Effects of L-Carnitine Administration on VO_2max and the Aerobic-Anaerobic Threshold in Normoxia and Acute Hypoxia." *European Journal of Applied Physiology* 60:1–6. 1990.